Essential Concepts of Animals in Sustainability

AF767570

This book compiles, discusses, and explains 126 key concepts relating to animals in sustainability in an easy-to-navigate A–Z format.

Essential Concepts of Animals in Sustainability is a roadmap of concepts that trace the connections of the planet with all its inhabitants – humans, non-human animals, other organisms, and ecosystems. This book provides an interdisciplinary introduction, drawing on the social, ecological, and physical sciences, to examine the place of animals as members of communities in relation to humans, other organisms, and the physical environment. Entries address the important role individual animals and species play in ecosystem management and environmental well-being, as well as addressing negative impacts on the environment, such as the methane produced by cattle. The book also focuses on the philosophical and ethical arguments that support protecting animals from human exploitation, including questions around emotion and sentience. Overall, this book provides students with a clear picture of the animal question in sustainability and will demonstrate that the ability to coexist is essential to the entire biophysical world.

This book will be of great interest to students and scholars of animal studies, zoology, conservation biology, environmental studies, ecosystem services, environmental humanities, and sustainability.

A–Z Guides for Environment and Sustainability

This series provides accessible, easy-to-navigate overviews of a range of different topics related to environment and sustainability. Following an A–Z format, each book contains entries which map out an important concept or term and illustrates how it connects more broadly to other ideas and disciplines. With related terms and further reading included alongside the entries, these innovative volumes will be of great interest to students and scholars learning, teaching and researching in this field.

Essential Concepts of Land Politics
An A–Z Guide
Saturnino M. Borras Jr. and Jennifer C. Franco

Essential Concepts for Implementing the Sustainable Development Goals
An A–Z Guide
Edited by Frank Biermann, Thomas Hickmann, Yi hyun Kang, Carole-Anne Sénit and Yixian Sun

Essential Concepts of Environmental Peacebuilding
An A-Z Guide
Edited by Carl Bruch, Rachel Weaver, Richard Matthew, Erika Weinthal, and Larry Swatuk

Essential Concepts of Science Communication
An A-Z Guide
Edited by Declan Fahy and Todd P. Newman

Essential Concepts of Animals in Sustainability
An A-Z Guide
Linda Kalof, Cameron Thomas Whitley and Thomas Dietz

For more information about this series, please visit: www.routledge.com/A-Z-Guides-for-Environment-and-Sustainability/book-series/AZES

Essential Concepts of Animals in Sustainability

An A-Z Guide

Linda Kalof, Cameron Thomas Whitley and Thomas Dietz

Routledge
Taylor & Francis Group
LONDON AND NEW YORK

Designed cover image "Critically Endangered Ploughshare Tortoise":
Tim Flach, photographer

First published 2027
by Routledge
4 Park Square, Milton Park, Abingdon, Oxon OX14 4RN

and by Routledge
605 Third Avenue, New York, NY 10158

Routledge is an imprint of the Taylor & Francis Group, an informa business

© 2027 Linda Kalof, Cameron Thomas Whitley and Thomas Dietz

The right of Linda Kalof, Cameron Thomas Whitley and Thomas Dietz to be identified as authors of this work has been asserted in accordance with sections 77 and 78 of the Copyright, Designs and Patents Act 1988.

All rights reserved. No part of this book may be reprinted or reproduced or utilised in any form or by any electronic, mechanical, or other means, now known or hereafter invented, including photocopying and recording, or in any information storage or retrieval system, without permission in writing from the publishers.

For Product Safety Concerns and Information please contact our EU representative GPSR@taylorandfrancis.com. Taylor & Francis Verlag GmbH, Kaufingerstraße 24, 80331 München, Germany.

Trademark notice: Product or corporate names may be trademarks or registered trademarks, and are used only for identification and explanation without intent to infringe.

British Library Cataloguing-in-Publication Data
A catalogue record for this book is available from the British Library

ISBN: 978-1-032-86568-3 (hbk)
ISBN: 978-1-032-86566-9 (pbk)
ISBN: 978-1-003-52814-2 (ebk)

DOI: 10.4324/9781003528142

Typeset in Optima
by Newgen Publishing UK

Contents

Introduction

Essential Concepts of Animals in Sustainability is a roadmap of 126 concepts that traces the connections of the planet with all its inhabitants – humans, non-human animals, other organisms, and ecosystems. Such conceptual clarity is essential for addressing the problems of and prospects for sustainability (Drury et al. 2023). However, clarity is challenging when so many disciplines, fields of study, and intellectual traditions must be engaged for an adequate understanding of the complex issues involved. For instance, differences in how disciplines conceptualize "animals," "nature," and "sustainability" can obscure shared concerns. These differences are not merely semantic but reflect deeper epistemological and normative divergences that shape how problems are defined and what solutions are considered. This book intends to facilitate communication across lines of scholarship and of engagement that are focused on animal studies and on sustainability discourse. Each of these two areas are already highly interdisciplinary and entwined. More integrated connections will yield synergies important for both understanding and action. Such synergies are particularly important in contexts where ethical considerations, ecological processes, and social institutions intersect in complex ways. By organizing key concepts at the intersection of animal studies and sustainability, this book provides a shared vocabulary useful for scholars, students, and practitioners seeking to better understand the importance of considering non-human animals in sustainability challenges and solutions.

Sustainability, in the sense of being able to continue to make use of a resource over time without depleting it, emerges in some of the earliest writing on resource management (Dietz 2023). Early conceptions of sustainability tended to center on the long-term maintenance of resource stocks, particularly in sectors such as forestry and fisheries, where continued use could deplete the capacity for regeneration. Recent interest in the idea flows mostly from what is commonly called the Brundtland Report on economic development and the environment. There sustainable development is defined as "development that meets the needs of the present without compromising the ability of future generations to meet their own needs" (World Commission on Environment and Development 1987, 43). This definition has led to extensive

DOI: 10.4324/9781003528142-1

analysis, including calls for a "science of sustainability" (Kates et al. 2001) and programs for action such as the elaboration of 17 specific Sustainable Development Goals to instantiate development under the auspices of the United Nations (2025). Despite their breadth, these goals unevenly incorporate non-human animals, often treating them as components of ecosystems rather than as subjects of concern. An animal studies perspective urges us to refine and expand these and other sustainability efforts to take more explicit and nuanced account of non-human animals (Torpman and Röcklinsberg 2021).

Such anthropocentric approaches consider non-human animals only to the extent that they meet human needs. Of course, since non-human animals are intertwined in much of human activity, even these anthropocentric views engage animals extensively and in meaningful ways. Yet such engagement rarely entails a fundamental reconsideration of the moral status of non-human animals. Recognizing how multifaceted the human and non-human animal world is, demands are increasing for the direct consideration of the well-being of non-human animals in sustainability analysis, including the Convention on Biological Diversity and the Sustainable Development Goals (Torpman and Röcklinsberg 2021; Verkuijil et al. 2025). These socio-cultural developments reflect a broader shift toward multispecies and more-than-human approaches in environmental scholarship.

Here we define sustainability as enhancing the well-being of humans and other species while minimizing harm to the environment (Dietz 2023). This definition deliberately broadens the scope of sustainability to include ethical considerations of non-human life. To operationalize this definition, we identify three key dimensions through which non-human animals are incorporated into sustainability. First, as in more anthropocentric definitions, human well-being depends very much on other species. Interactions with other species, both domestic and "wild" animals, are important sources of enjoyment and comfort for many humans, and the use of animals remains a key part of human livelihoods in many places. These interdependencies highlight the extent to which human well-being is embedded in broader multispecies relationships.

Second, the well-being of other species is an explicit criterion for sustainability. This is consistent with many ethical theories in philosophy as well as many Indigenous traditions that argue for the consideration of the well-being of non-human species in approaches to global viability. Such perspectives challenge narrow approaches by emphasizing moral obligations that extend beyond human communities.

Third, the environment, a broad term used to cover ecosystems from the smallest scale such as tidal pools and microbial mats to the biosphere, includes non-human species. The dynamics of environmental change, from micro-ecosystems to large-scale global change processes like extinctions and climate change all impact non-human animals who in turn shape those dynamics. Recognizing animals as active participants in ecological processes underscores their role as agents in environmental systems.

While approaches to sustainability require careful consideration of non-human animals, the scholarship in sustainability and animal studies has only rarely intersected, with limited engagement between the two. This paucity is ironic since human cultural and biological evolution has always been in interplay with other species (Dietz 2023). Thus, the lack of integration has limited the ability of both fields to fully account for the complexity of human-animal-environment relationships. Because humans and other animals have always lived in co-constituted, collaborative worlds (Despret 2013; Haraway 2007; Porcher 2017), the field of animal studies is an interdisciplinary endeavor to understand the multilayered human-animal relationship in both historical and contemporary contexts (Kalof 2017). Thus, animal studies encompasses a wide range of disciplines, including cultural studies, legal studies, philosophy, sociology, psychology, history, and zooarchaeology. Intellectual struggles with the "animal question" began in the 1970s and have now assumed a priority position in the discourse of politics, ethics, public policy, and law because of the widespread recognition of animal exploitation, the degradation of the natural world, the loss of animal habitat and species extinction, and our need to coexist with other animals in urban, rural, and natural contexts (Kalof 2017).

Animal studies makes a substantial contribution to sustainability by extending ethical concerns beyond narrow human interests to include the lived experiences, well-being, sentience, and cognition of non-human animals and the ecosystems they inhabit. By foregrounding empathy for animals, this perspective draws attention to the deeply entangled relationships through which human and non-human lives mutually shape ecological futures (Whitley et al. 2025). In doing so, it reframes sustainability as a relational process grounded in interdependence rather than as a purely technical problem. In this sense, sustainability is not solely a policy or scientific challenge but an ethical commitment that implicates all life on the planet. Further, animal studies helps to unpack the complex coevolution of humans and other species, belying overly simplistic notions of those interactions. Indeed, recent conceptions of life on Earth and its evolution note that each individual animal is itself an ecosystem of both its own cells and comparable numbers of bacteria and viruses. Thus, a realistic perspective of life on Earth does not draw a rigid boundary between animals and other species and views the planet as a nested series of ecosystems ranging from one individual through ecosystems to the biosphere (Leonard et al. 2024). This perspective challenges conventional boundaries between animals (human and non-human), species, microorganisms, and environments, emphasizing the continuity of life across all scales. Animal studies has much to offer the evolving discourse on sustainability. In turn, many of the concerns that motivate the animal studies literature can benefit from using the broad perspective of sustainability to consider contexts and connections that might otherwise be underappreciated. We hope that with this volume more communication and even integration between animal studies and sustainability will be facilitated and richer and more robust perspectives will emerge.

In this volume we take an interdisciplinary approach to mapping the key concepts in the discourse of animals in sustainability. This mapping is intended to make visible the connections among concepts that are often treated in isolation. Our work draws on the social, ecological, and physical sciences, as well as the humanities and Indigenous knowledge, to examine the place of animals as members of communities with humans, other organisms, and the physical environment. Individual animals, species, and assemblages such as ecosystems and landscapes all need to be considered in realistic and comprehensive sustainability objectives. Their well-being contributes to sustainability even when it is defined in a narrow anthropocentric way, and they are worthy of ethical consideration in themselves. Thus, in framing the role of animals in sustainability we include essential concepts and theories such as island biogeography, ecological restoration, and threatened ecosystems. Additionally, entries cover some of the most pernicious human-animal entanglements that have negative impacts on the environment, such as livestock production which is a major source of biodiversity loss throughout the world.

Each entry provides brief, well-documented definitions, overviews, and examples, of concepts and issues. Throughout each entry, we make connections to related concepts to encourage readers to explore the network of interconnections that is the essence of tracing the theme of animals in sustainability discourse. At the same time, the conceptual terrain we map here should be understood as a starting point to a rapidly evolving field, characterized by the continual emergence of new terms, frameworks, and lines of research. Although this volume brings together 126 key concepts, we do not claim to capture the full scope of relevant ideas. After reading an entry you won't be an expert on the topic. Rather, the goal is to provide a conceptual orientation and a foundation for more advanced inquiry, and you will have a sense of what a term means, why it is important for understanding the linkages of animals to sustainability, and how it connects to other ideas and concepts. Our hope is to give you an accessible starting point for learning more, with enough perspective to help you decide which topics to pursue further and references that serve as a first step for that pursuit.

The overarching message of this book is that we live in an ecological community of living beings all dependent on each other and the resources of the planet to survive, so the ability to coexist is essential (Kalof 2027). To adequately address sustainability, animals must be central to scientific and ethical frameworks that expand the scope of concern to include the well-being of all species and ecosystems on the planet. Recognizing this interdependence is not simply an analytical exercise but a necessary step toward developing more equitable forms of coexistence.

References

Despret, Vinciane. 2013. "Responding Bodies and Partial Affinities in Human-animal Worlds. *Theory, Culture & Society* 40 (7/8): 51–76.

Dietz, Thomas. 2023. *Decisions for Sustainability: Facts and Values*. Cambridge University Press.

Drury, Matt, Janet Fuller, and John Hoeks. 2023. "Embedding Animals Within a Definition of Sustainability." *Sustainability Science* 18:1925–1938.

Haraway, Donna. 2007. *When Species Meet*. University of Minnesota Press.

Kalof, Linda. 2017. "Introduction." In *The Oxford Handbook of Animal Studies*, edited by Linda Kalof. Oxford University Press.

Kalof, Linda. 2027. *Speaking of Real Animals and Their History: From the Deep Past to the Present*. Michigan State University Press.

Kates, Robert W., William C. Clark, Robert Corell, et al. 2001. "Sustainability Science." *Science* 292 (5517): 641–642.

Leonard, Aoife, Javier Abalos, Titus Adhola, et al. 2024. "A Global Initiative for Ecological and Evolutionary Hologenomics." *Trends in Ecology & Evolution* 39 (7): 616–620. https://doi.org/10.1016/j.tree.2024.03.005.

Porcher, Jocelyn. 2017. *The Ethics of Animal Labor: A Collaborative Utopia*. Palgrave.

Torpman, Ollie, and Helena Röcklinsberg. 2021. "Reinterpreting the SDGs: Taking Animals into Direct Consideration." *Sustainability* 13: 843.

United Nations. 2025. *The Sustainable Development Goals Report 2025*. United Nations. https://unstats.un.org/sdgs/report/2025/.

Verkuijil, Cleo, Jeff Sebo, Adalene Minelli, and Katherine Browne. 2025. *Integrating Animal Health and Welfare into the 2030 Agenda and Beyond*. Stockholm Environment Institute and Center for Environmental and Animal Protection.

Whitley, Cameron T., Linda Kalof, L.C. Urquhart, Nate Tatem, Melissa Mair, Katya Ankoudinova, Ingrid Haight, et al. 2025. "Photo Portraiture Enhances Empathy for Birds with Potential Benefits for Conservation and Sustainability." *Sustainability* 17 (19): 8833.

World Commission on Environment and Development. 1987. *Our Common Future*. Oxford University Press.

Author Biographies

Linda Kalof is Professor Emerita of Sociology and Environmental Science & Policy at Michigan State University. She uses an interdisciplinary lens to study the cultural and natural histories of animals and the role of animals in environmental sustainability. She has published 14 books, including *Looking at Animals in Human History* (Reaktion 2007), *A Cultural History of Animals in Antiquity* (Palgrave, 2007), *The Oxford Handbook of Animal Studies* (Oxford University Press 2017), *The Animals Reader, Second Edition* (Routledge 2022), and *Speaking of Real Animals and Their History: From the Deep Past to the Present* (Michigan State University Press 2027).

Cameron Thomas Whitley is Professor of Sociology at Western Washington University. With over six dozen publications, his research examines empathy for non-human animals, inclusive approaches to environmental justice, human-animal relationships in climate-induced disasters, and representations of animals with disabilities, with particular attention to care, vulnerability, and multispecies inequality. In 2023, he was awarded a five-year National Science Foundation Early CAREER grant to collaborate with accredited zoos and aquariums on research examining how animal imagery and storytelling elicit emotional responses and foster conservation-oriented behavior.

Thomas Dietz is University Distinguished Professor Emerit and Founding Director of the Environmental Science and Policy Program at Michigan State University. He studies human ecology, decision making for sustainability, and the human dimensions of environmental change. His recent publications include *Decisions for Sustainability: Facts and Values* (Cambridge University Press 2023). In 2024 he was elected to the US National Academy of Sciences.

A

Abolitionism

In the context of animal ethics, animal abolitionism is a philosophical and political approach that seeks to end the institutionalized use of non-human animals rather than merely reforming existing systems to increase welfare (→ Welfare). This perspective is rooted in debates about sentience, moral equality, and rights (→ Sentience) (→ Rights). Abolitionism rejects the idea that animals can be ethically used as property, resources, or commodities, regardless of how "humanely" they are treated (→ Property). Contemporary scholarship highlights that abolitionism is less about incremental welfare improvements and more about dismantling social structures that normalize animal exploitation in food systems, research, entertainment, and labor (Giroux 2024). Abolitionist theory challenges deep-seated cultural assumptions about human entitlements to animal bodies and lives.

Animal abolitionism is the basis of prominent critiques of organizations that do animal research, have captive wildlife, or are involved in livestock production (→ Animal Research) (→ Captive Wild Animals) (→ Zoos and Aquariums) (→ Livestock Production) (Vinnari and Vinnari 2022; Campbell 2013; Francione and Garner 2010). Animals such as pigs, cows, and chickens, are bred, confined, and slaughtered on a massive scale, despite extensive evidence demonstrating their cognitive, emotional, and social capacities (→ Animal Languages) (→ Animal Cultures) (→ Cognition) (→ Learning for Survival). While welfare reforms may improve conditions, abolitionist scholars argue that these reforms do not address the core ethical problem of animals as property, whose interests come second to human economic demands (Favre 2009; Francione 2012). For example, cage-free egg production eliminates battery cages, but still involves practices like chick culling and the restriction of movement. Abolitionists contend that such systems preserve exploitation while presenting an illusion of ethical progress. This example illustrates the central claim of abolitionism that true justice for animals requires ending their instrumental use.

Abolitionism intersects with sustainability in meaningful ways. Industrial animal agriculture is a leading driver of greenhouse gas emissions, land degradation, water pollution, and biodiversity loss. Abolitionist scholars argue that shifting away from animal-based food systems and promoting vegetarianism

DOI: 10.4324/9781003528142-2

and veganism (→ Veganism) (→ Vegetarianism) would reduce environmental pressures while recognizing the intrinsic value and ecological roles of non-human animals. By challenging speciesist assumptions and the economic structures that rely on mass animal exploitation (→ Speciesism), abolitionism encourages the reimagining of food systems, conservation strategies, and human-animal relationships. From this perspective, sustainability is not only about protecting ecosystems, but also about transforming the moral foundations that shape how societies relate to non-human animals.

References

Campbell, Kathryn M. 2013. "Zoos as Prisons: The Role of Law and the Case for Abolition." *Mid-Atlantic Journal on Law and Public Policy* 2: 53.

Favre, David. 2009. "Living Property: A New Status for Animals within the Legal System." *Marquette Law Review* 93: 1021.

Francione, Gary. 2012. *Animals Property & the Law*. Temple University Press.

Francione, Gary Lawrence, and Robert Garner. 2010. *The Animal Rights Debate: Abolition or Regulation?* Columbia University Press.

Valéry Giroux. 2024. "Abolitionism." In *The Plant-Based and Vegan Handbook: Psychological and Multidisciplinary Perspectives*, edited by Yanoula Arthanassakis, Renan Larue, and William O'Donohue, 69–87. Springer.

Vinnari, Eija, and Markus Vinnari. 2022. "Making the Invisibles Visible: Including Animals in Sustainability (and) Accounting." *Critical Perspectives on Accounting* 82: 102324.

Adaptation

Adaptation is a central concept in the theory of evolution (→ Evolution). Early evolutionary theory argued that some individuals would be more successful at surviving and reproducing because they were better able to cope with the environment in which they lived, including their interactions with other species. The genes that shape those individuals would be more prevalent in the next generation, and over time a species would "adapt" to the environment by having genes that produced individuals well suited to the environment. Some genetically determined traits would be favored over others, and over generations the species would change, would evolve – this is sometimes called the survival of the fittest.

While such processes are still considered very important, current ideas about adaptation are more nuanced (Laland et al. 2024; Gould 2002). Animals move across ecosystems and thus are shaped by more than one environment and can to some degree choose their environment. The traits that are best for one environment may differ from the traits best for another, making adaptation complex. Some adaptation comes from the biological development of the individual – two animals with identical genomes can develop to be somewhat different because of processes that occur after conception. Animals learn from experience which can lead to adaptation, and animals have culture – they communicate with each other and observe and learn from each other (→ Learning for Survival) (→ Animal Cultures) (→ Cultural Transmission), leading to cultural evolution, an important force in the adaptation of many species (→ Evolution).

Modern evolutionary theory invokes the idea of an adaptive landscape. High points on this theoretical landscape correspond to a good fit to the environment, while valleys correspond to traits and behaviors that challenge an individual's ability to survive and reproduce. The processes that shape adaptation tend to push species and individuals from the valleys to the peaks. But animals are also constantly reshaping the landscape, leading to the idea of a constructed adaptive landscape (McLaughlin and Dietz 2008). For example, many species actively reshape the physical structure of their environments, from corals who form reefs to beavers who construct dams and ponds (→ Ecosystem Engineers).

Some animals adapt to unusual environmental conditions, such as disasters or changes in the density of human populations. For example, the gray wolf population in the Chornobyl Exclusion Zone is adapting to chronic radiation exposure, suggesting that the wolves are developing cancer-resistant or cancer-resilient genes (Love et al. 2026). And on the campus of the University of California at Los Angeles, the dark-eyed junco, a common songbird who has lived alongside students for generations, developed a change in the shape of their beaks during the pandemic campus shutdown of 2020. The normal shape of the junco beak is short and stubby, best suited for scavenging a diet of human leftovers such as chips, bread, and cookies. However, the birds born during the quiet years of the shutdown did not have access to anthropogenic food and because they fed on food typical of wild areas such as insects, seeds, and grasses, they developed distinctly thin, long beaks similar to their wild conspecifics (Marzluff 2024).

Global environmental change (→ Global Environmental Change) is rapidly transforming many environments. These changes often lead to shifts in the adaptive landscape. In some cases, species can move quickly to the new peaks by changing their adaptive strategy. But adaptations based on slow processes such as genetic change may not be able to keep up with environmental change, and that can lead to extinction and biodiversity loss (→ Extinction) (→ Defaunation). Thus, understanding adaptive processes is crucial for addressing sustainability.

References

Gould, Stephen Jay. 2002. *The Structure of Evolutionary Theory*. Harvard University Press.

Laland, Kevin N., David Andrews, Tobias Uller, Nathalie Feiner, Marcus W. Feldman, and Scott F. Gilbert. 2024. *Evolution Evolving: The Developmental Origins of Adaptation and Biodiversity*. Princeton University Press.

Love, Cara N., Stacey L. Lance, Thomas G. Hinton, Nicolas Rochette, James C. Beasley, Dmitry Shamovich, Michael E. Byrne, Brian Nadel, Sarah C. Webster, and Shane C. Campbell-Staton. 2026. "Signatures of Radiation-Induced Stress and Putative Selection on Immune Targets in Chornobyl Wolves." *Molecular Ecology* 35(9): e70308. doi:10.1111/mec.70308.
Marzluff, John M. 2024. "Human Action Shapes the Beaks of Backyard Birds." *Proceedings of the National Academy of Sciences* 123 (4): e2535458123. https://doi.org/10.1073/pnas.2535458123.
McLaughlin, Paul, and Thomas Dietz. 2008. "Structure, Agency and Environment: Toward an Integrated Perspective on Vulnerability." *Global Environmental Change* 18 (1): 99–111.

Advocacy

Advocacy refers to the strategic efforts taken by individuals, organizations, or communities to influence public opinion, institutional practices, or policy decisions in the pursuit of social or environmental change. Advocacy for animals involves a range of activities such as public education, protest, lobbying, coalition building, and participation in governance processes, that aim to amplify concern for non-human animals and policy changes to support non-human animals. Scholars note that effective animal advocacy typically combines evidence-based research, moral argumentation, and community mobilization to challenge existing power structures and promote more equitable outcomes for humans and non-human animals (Rudy 2011). Animal advocates often highlight ethical concerns about sentience, welfare, and rights while questioning the norms that legitimize exploitation (→ Rights) (→ Sentience) (→ Welfare) (Fitzgerald 2018).

Animal advocacy demonstrates how these efforts shape public debates about non-human animals' moral and legal status. As an example, there is a growing movement to restrict or abolish intensive animal confinement systems such as gestation crates for pigs and battery cages for hens. Advocates draw on scientific studies documenting animals' cognitive capacities, pain perception, and social needs to argue that industrial systems violate basic standards of humane treatment (→ Cognition) (→ Sentience). Much of this work is in the realm of comparative psychology. Marino and colleagues (2015) highlight the cognitive abilities of pigs to suggest that animal research involving pigs should be refined and limited because pigs are social, self-aware, show emotion, and have personalities (→ Animal Research) (→ Agency). Their work has contributed to concrete changes, including corporate commitments to cage-free supply chains and state-level legislation improving welfare protections. At the same time, rights-based advocacy challenges deeper assumptions by arguing that non-human animals possess inherent value that should not be overridden for convenience or profit (Regan 2004). These campaigns illustrate how advocacy combines ethical claims, empirical evidence, and public engagement to shift norms and policy landscapes. Advocates are using their agency to shape cultural evolution toward a more sustainable and just society by changing both norms and legal rules.

Animal advocacy is central to sustainability because many environmental challenges such as habitat loss, climate change, pollution, and biodiversity decline directly affect non-human animals (→ Habitat Loss) (→ Climate Change) (→ Pollution) (→ Defaunation) (→ Extinction) (Broad 2018). Industrial livestock production, deforestation, and ocean degradation threaten species survival and undermine ecosystem resilience. Advocates working at the intersection of animal protection and environmental justice argue that sustainable futures require transforming food systems, strengthening habitat protections, and recognizing animals as ecological participants rather than commodities (→ Livestock Production) (→ Environmental Justice) (Sebo 2021). By connecting animal welfare, rights, and environmental justice, advocacy helps promote policies and practices that support both ecological integrity and the well-being of non-human animals.

References

Broad, Garrett M. 2018. "Effective Animal Advocacy: Effective Altruism, the Social Economy, and the Animal Protection Movement." *Agriculture and Human Values* 35 (4):777–789.

Fitzgerald, Amy J. 2018. *Animal Advocacy and Environmentalism: Understanding and Bridging the Divide*. Wiley.

Marino, Lori, and Christina M. Colvin. 2015. "Thinking Pigs: A Comparative Review of Cognition, Emotion, and Personality in *Sus domesticus*." *International Journal of Comparative Psychology* 28 (1): 1–26. https://doi.org/10.46867/ijcp.2015.28.00.04.

Regan, Tom. 2004. *The Case for Animal Rights*. University of California Press.

Rudy, Kathy. 2011. *Loving Animals: Toward a New Animal Advocacy*. University of Minnesota Press.

Sebo, Jeff. 2021. "Animals and Climate Change." In *Philosophy and Climate Change*, edited by Mark Budolfson, Tristram McPherson, and David Plunkett, 42–66. Oxford University Press. https://doi.org/10.1093/oso/9780198796282.003.0003

Agency

Agency is the ability of an animal to make conscious choices that have an impact on their lives and on the lives of others and the environment (Dietz and Burns 1992). Although some strains of human exceptionalism deny it (→ Anthropocentrism), the consensus from scientific research or simply careful observation makes it clear that many non-human species have the potential for agency if they are permitted to make choices. Under most ethical theories, providing animals opportunities to exercise their agency is considered desirable or even essential (→ Capabilities Ethics) (→ Ecodemocracy) (→ Utilitarianism).

Even when the exercise of agency by animals is considered desirable, determining an appropriate course of action can be challenging.

For example, many zoos and aquaria (→ Captive Wild Animals) have breeding programs intended to perpetuate the genetic diversity of species with low population numbers. Such programs are most effective when mating choice is managed to minimize inbreeding, enhance genetic diversity, and preserve rare genetic lines. However, such planning obstructs animal agency in reproductive choice (Norman and Brando 2024). Recommendations for the care of animals who are used in research increasingly address the issue of providing the animals with some scope for agency (→ Animal Research) (Animal Welfare Institute, n.d.)

There is a growing realization that domesticated animals are active collaborators in the work that goes on in agriculture (→ Livestock Production) (Porcher and Schmitt 2012). In emerging systems that automate milking, cows can choose when they want to be milked, and in general this ability to exert agency improves animal welfare (Gislon et al. 2024). Animals also show agency when working in entertainment, law enforcement, as care workers, and in many other roles (→ Disability) (Porcher and Estebanez 2019; Hribal 2007). In general, most contemporary discussions take account of agency as a key factor in welfare (→ Welfare). Views of sustainability that take into consideration the well-being of non-human species thus must take account of agency. Even anthropocentric views must consider the agency of animals in many processes such as agriculture that are central to sustainability (→ Anthropocentrism) (→ Livestock Production).

References

Animal Welfare Institute. n.d. "Non-Human Primates." Accessed November 28, 2025. https://awionline.org/content/non-human-primates.

Dietz, Thomas, and Tom R. Burns. 1992. "Human Agency and the Evolutionary Dynamics of Culture." *Acta Sociologica* 35: 187–200.

Gislon, Giulia, Luciana Bava, Maddalena Zucali, Alberto Tamburini, and Anna Sandrucci. 2024. "Unlocking Insights: Text Mining Analysis on the Health, Welfare, and Behavior of Cows in Automated Milking Systems." *Journal of Animal Science* 102 (January): skae159. https://doi.org/10.1093/jas/skae159.

Hribal, Jason. 2007. "Animals, Agency, and Class: Writing the History of Animals from Below." *Human Ecology Review* 14 (1): 101–112.

Norman, Max, and Sabrina Brando. 2024. "The Concept of Agency, Animal Wellbeing, and the Practical Realities of Ex Situ Breeding Programs in Zoos and Aquariums." *Journal of Zoological and Botanical Gardens* 5 (4): 563–578. https://doi.org/10.3390/jzbg5040038.

Porcher, Jocelyne, and Jean Estebanez, eds. 2019. *Animal Labor: A New Perspective on Human-Animal Relations*. Transcript Verlag.

Porcher, Jocelyne, and Tiphaine Schmitt. 2012. "Dairy Cows: Workers in the Shadows?" *Society & Animals* 20 (1): 39–60. https://doi.org/10.1163/156853012X614350.

Altruism

Altruism is the action of helping others, even if it costs you something. Frans de Waal (2008, 280) writes that "an action is called 'altruistic' if it benefits a recipient at a cost to the actor." Many animals, including humans, often make big sacrifices for their young and other relatives, and some species also help non-relatives and even members of other species (Clutton-Brock 2009). For example, research suggests that the humpback whales' frequent defense of other species who are being attacked by killer whales is an example of inter-specific altruism (Pitman et al. 2017).

Since being altruistic seems like a disadvantage to the altruist, evolutionary theorists have tried to understand why it occurs (Sober and Wilson 1999). One explanation is genetic kin selection where evolution favors those who act altruistically toward others who share their genes. Kin selection shows why ants, bees, and other eusocial insects help their colony – all members are very closely related. Reciprocal altruism happens when you help someone hoping they will help you in return. In group selection, everyone shares the same fate. So, when you act for the group's good, you benefit as well. For example, when a robin, crow, or bluejay spots a predator and makes a warning call, she draws attention to herself but warns the other members of the flock, including both kin and non-kin.

These ideas assume that genes play a primary role in shaping an animal's actions. The theory of cultural evolution says that learning from others is important (→ Learning for Survival) (Boyd and Richerson 2024). Cultural evolution can lead to altruism for those who aren't your genetic relatives, especially for those you see as part of your social group. If everyone acts selfishly, it becomes hard to coordinate actions that benefit all. Yet such coordination is essential for sustainability. Researchers refer to this as the tragedy of the commons or the problem of collective action (→ Commons Dilemma) (Ostrom et al. 2002). Altruism is important in taking group action for the common good. For example, research in social psychology finds that being altruistic leads to positive actions toward the environment (Dietz and Whitley 2018).

References

Boyd, Robert, and Peter J. Richerson. 2024. "Cultural Evolution: Where We Have Been and Where We Are Going (Maybe)." *Proceedings of the National Academy of Sciences* 121 (48): 1–10. https://doi.org/10.1073/pnas.2322879121.

Clutton-Brock, Tim. 2009. "Cooperation between Non-Kin in Animal Societies." *Nature* 462 (7269): 51–57. https://doi.org/10.1038/nature08366.

de Waal, Frans B.M. 2008. "Putting the Altruism Back into Altruism: The Evolution of Empathy." *Annual Review of Psychology* 59: 279–300.

Dietz, Thomas, and Cameron T. Whitley. 2018. "Inequality, Decisions, and Altruism." *Sociology of Development* 4 (3): 282–303.
Ostrom, Elinor, Thomas Dietz, Nives Dolsak, Paul C. Stern, Susan Stonich, and Elke Weber. 2002. *The Drama of the Commons*. National Academy Press.
Pitman, Robert L., Volker B. Deecke, Christine M. Gabriele, et al. 2017. "Humpback Whales Interfering When Mammal-Eating Killer Whales Attack Other Species: Mobbing Behavior and Interspecific Altruism?" *Marine Mammal Science* 33 (1): 7–58. https://doi.org/10.1111/mms.12343.
Sober, Elliott, and David Sloan Wilson. 1999. *Unto Others: The Evolution and Psychology of Unselfish Behavior*. Harvard University Press.

Animal Ambassadors

Animal ambassadors are individual animals housed or cared for by zoos, aquariums, nature centers, rehabilitation facilities or education organizations who are presented to the public to enhance learning and often promote conservation behaviors (→ Wildlife Tourism). Presentation of these animals occurs in the form of talks, demonstrations, and closely facilitated encounters. Animal ambassadors fall within Animal Interaction Programming (AIPs). Broadly, AIP is defined as planned animal-visitor interactions (AVIs) for entertainment or educational purposes. While always planned, AVIs can be either direct or indirect, and scripted or unscripted. Animal ambassador encounter programs (AAEPs) are one type of AIP focused exclusively on education and pro-environmental behavior enhancement. The use of animal ambassadors is considered a conservation education (CE) tactic designed to increase awareness about biodiversity and encourage pro-environmental behaviors by engaging people in controlled animal interactions with animals that they would not likely encounter, or that are heavily feared or vilified. Animal ambassadors can be from a variety of species and include both vertebrates and invertebrates. As far back as the 1960s, zookeepers would occasionally appear with individual animals on popular television programs such as Johnny Carson's Tonight Show, but the modern use of animal ambassadors is attributed to Jack Hanna, US zoo-keeper and conservationist, and his show from the 1980s that focused on presenting animals as ambassadors to connect people with specific species and environmental issues.

The efficacy of AAEPs is often debated. For instance, in a quasi-experimental study, the use of penguin animal ambassadors led to improved knowledge and higher behavior intentions (Clifford-Clarke et al. 2021). Counter to this, Kirsch et al. (2025) found that animal encounters with a parrot or python did not impact conservation attitudes. It is possible that differences in efficacy could be due to instructional variations in program design, messaging, species, and training among other features. Spooner et al. (2021) asserts that the limited and mixed findings warrant a deeper discussion about the value of AAEPs and the welfare needs of animal ambassadors.

There is a lack of data and understanding about the impact of AAEPs on animal welfare (Cooper-Rogers et al. 2026). Much of the work to assess welfare among animal ambassadors focuses on mammalian species with only limited welfare measures used (→ Welfare) (Cooper-Rogers et al. 2026). Better standardized and species-specific behavior and psychological welfare indicators are needed to assess animal well-being among animals used in educational programming (Martin et al. 2024). Animal ambassadors may also assume roles as animal celebrities as organizations, media outlets, and online public forums shape and distribute their stories to raise money for care and conservation initiatives. This can lead to expanded concerns over welfare. For example, Fiona the Hippo, born prematurely at the Cincinnati Zoo in 2017, quickly became an international sensation. Media coverage often framed her story in human-like terms, allowing audiences to emotionally connect with her recovery. That widespread attention translated into tangible conservation outcomes, generating more than $500,000 in donations directed toward hippo conservation efforts (Saker 2018). Although Fiona the Hippo is not the type of animal typically described as an ambassador, given that she cannot be easily transported or handled, she nonetheless fulfills many ambassador functions. She features prominently in educational talks and demonstrations at the zoo, and her image has been widely used in outreach and awareness campaigns designed to engage the public with conservation issues.

Not all animal ambassadors live in zoos, aquariums, or other managed environments. Increasingly, wildlife living freely in their natural habitats are becoming public ambassadors through digital-monitoring technologies that allow people to observe them remotely. A well-known example is the bald eagle pair Jackie and Shadow who nest near Big Bear Lake in California's San Bernardino National Forest. Wildlife cameras installed near their nest by the nonprofit organizations Friends of Big Bear Valley stream live footage of the pair's daily activities, allowing viewers around the world to watch as they build their nest, incubate eggs, and raise their young. Jackie first appeared at the nest site in the mid-2010s, and Shadow later joined her as a mate. Since then, their seasonal nesting cycles have attracted a large and devoted online audience (Friends of Big Bear Valley 2026). During nesting seasons, tens of thousands of viewers regularly turn into the livestream, forming a global community that follows the pair's successes and challenges. This visibility has translated into tangible conservation support. Within a relatively short period, public enthusiasm surrounding Jackie and Shadow have contributed to fundraising efforts that, as of July 2026, have raised roughly $5 million to help prevent nearby development that threatens the habitat surrounding their nesting area (Friends of Big Bear Valley 2026).

The widely followed stories of Fiona, a hippopotamus born under human care, and Jackie and Shadow, a pair of bald eagles nesting in the wild, demonstrate how animals observed even through cameras can become influential ambassadors for conservation. Whether in managed settings or natural habitats, these mediated encounters allow people to witness the lives of individual animals in ways that were rarely possible in earlier generations. As audiences follow these animals' daily routines, challenges, and milestones, many develop a sense of familiarity and concern that extends beyond simple curiosity for animals they have never met and never will meet.

Animal ambassadors are often used in informal learning environments and free-choice learning institutions as high-impact interpretive tools that can contribute to sustainability by making abstract and distant environmental problems, such as pesticide use, plastic pollution, palm oil use, and climate change, more emotionally and locally tangible (→ Climate Change) (→ Pollution). There is a growing body of literature engaged with how to better utilize AAEPs to enhance attitudes and behaviors, while also adhering to welfare concerns (→ Welfare). If done correctly, AAEPs can be powerful tools in situating and localizing the impact of environmental problems and needed behavioral change.

References

Clifford-Clarke, Megan Marie, Katherine Whitehouse-Tedd, and Clare Frances Ellis. 2021. "Conservation Education Impacts of Animal Ambassadors in Zoos." *Journal of Zoological and Botanical Gardens* 3 (1): 1–18. https://doi.org/10.3390/jzbg3010001

Cooper-Rogers, Bridget, Paige Klingner, Alex Whittaker, Bec Forder, and Eduardo J. Fernandez. 2026. "Program Animal Welfare: A Scoping Review of Animal Interaction Programs and Their Welfare Impact on Zoo-Housed Animals." *Applied Animal Behaviour Science* 297: 106919. https://doi.org/10.1016/j.applanim.2026.106919

Friends of Big Bear Valley. 2026. "Jackie and Shadow Need All of You to Be Their Heroes." Accessed March 22, 2026. https://savemooncamp.org/.

Kirsch, Daniel G., Anna C. MacPherson, Caterina N. Meyers, Sarah L. Bowser, and Sara M. Kross. 2025. "Ambassador Animals Do Not Have a Clear Effect on Visitor Conservation Knowledge and Attitudes Toward Exotic Pets at a Zoo Exhibit." *Zoo Biology* 44 (1): 36–48. https://doi.org/10.1002/zoo.21872

Martin, Steve, Grey Stafford, and David S. Miller. 2024. "A Reexamination of the Relationship between Training Practices and Welfare in the Management of Ambassador Animals." *Animals* 14 (5): 736. https://doi.org/10.3390/ani14050736

Saker, Anne. 2018. "Fiona Inc.: Nearly $500K to Cincinnati Zoo from Sales of Hippo Books, Beer and Ice Cream." *Cincinnati The Enquirer* (Cincinnati, OH), January 24. www.cincinnati.com/story/news/2018/01/03/fiona-inc-nearly-500-k-cincinnati-zoo-sales-hippo-books-beer-and-ice-cream/999850001/?gnt-cfr=1&gca-cat=p&gca-uir=true&gca-epti=z1183xxu002432e1183xxv002932&gca-ft=233&gca-ds=sophi.

Spooner, Sarah L., Mark J. Farnworth, Samantha J. Ward, and Katherine M. Whitehouse-Tedd. 2021. "Conservation Education: Are Zoo Animals Effective Ambassadors and Is There Any Cost to Their Welfare?" *Journal of Zoological and Botanical Gardens* 2 (1): 41–65. https://doi.org/10.3390/jzbg2010004.

Animal Cultures

Culture is group-typical behaviors shared by community members that rely on socially learned and transmitted information (Hersh et al. 2025). While all animals learn to some degree, "a capacity for social learning is merely a prerequisite for culture" (→ Learning for Survival) (Laland and Janik 2006, 545). Culture in animals is demonstrated by differentiating socially contingent learning from asocial learning, such as when individuals newly introduced to an ecosystem adopt the same behaviors as the established residents (Laland and Janik 2006). Some well-documented examples of cultural traditions in animals include the migratory routes of ungulates (Jesmer et al. 2025) and reef fish (Brown and Webster 2025), stone tool use in wild nonhuman primates (Kalan and Luncz 2025), and vocalizations in wild birds (→ Cultural Transmission) (Crates et al. 2025).

The cultural traditions of cetaceans are particularly well documented. Toothed whales, dolphins, and porpoises have socially learned foraging tactics, such as dolphins' use of mangroves as barriers to trap fish and their use of marine sponges to oust bottom-dwelling fish from the ocean floor (Hersh et al. 2025). The cultural transmission of specialized foraging tactics in odontocetes is from mother to offspring during the first year of life, a vertical transmission of learned behavior that highlights "female-based sociality as a cornerstone of cetacean societies and the substrate for cultural transmission" (Hersh et al. 2025, 7). Male humpback whales also have novel foraging strategies, in addition to complex song patterns that are socially transmitted in close acoustic contact (such as shared feeding grounds or migration routes), allowing for the horizontal transmission of complex song arrangements to be learned between populations (Allen et al. 2022).

Animal cultures are disappearing worldwide. Behaviors can disappear because of 1) the loss of old individuals with specific behaviors such as bird song dialects or accumulated wisdom such as migratory routes or food sources, or 2) human-induced environmental changes, "drift" in dwindling populations as the number of models to be copied decreases, or species extinction (Caro and Sherman 2012). Island populations are particularly vulnerable to human-caused extinction, such as the Steller's sea cow on the Commander Islands who was exterminated by human hunting within 27 years of their discovery in the eighteenth century (Roopnarine et al. 2022) and the Christmas Island shrew who was declared extinct in October 2025 primarily from predation by feral cats and rats (→ Feral Animals) (→ Extinction) (IUCN 2025).

Caro and Sherman (2012) provide examples of behavioral disappearance caused by humans: loss of prey naiveté in African ungulates from human hunting; loss of predator recognition in moose from the extirpation of wolves; loss of migratory routes for elephants, bison, springbok, Malayan sun bears, and bearded pigs from anthropogenic habitat alterations and fragmentation; and loss of song dialects in White-crowned sparrows from urban noise (→ Ecology of Fear) (→ Habitat Loss) (→ Pollution). While humans have caused the disappearance of many animal behaviors, human intervention can be used to restore animal behavior in the wild, such as training marsupials prior to their release to recognize nonnative predators, teaching whooping cranes to undergo annual migrations following an ultralight plane, removing dams to allow salmon to migrate upstream, removing fences that prevent mammal movements, and teaching black-footed ferrets to hunt prey (Caro and Sherman 2012). In an experiment to reintroduce wild song culture to an endangered songbird, zoo-bred juvenile honeyeater songbirds were taught wild songs recorded from wild-origin males (Appleby et al. 2026). The proportion of juveniles who learned the wild song was 42 percent after three years, and the entire population is predicted to sing the wild song within two years; the wild song taught to zoo-bred birds disappeared from the wild because of declining populations, making the zoo population the only remaining reservoir of traditional song culture (Appleby et al. 2026).

An emphasis on social learning in animal populations also highlights the importance of old, wise individuals as repositories of accumulated knowledge (→ Learning for Survival). The protection of old individuals is crucial for the preservation of social groups, but the traditional conservation approach prioritizes younger individuals for their potential to reproduce (Brakes et al. 2019). That conservation focus along with recreational bowfishing has contributed to the decline of fish such as the Bigmouth Buffalo who, at 112 years of age, is the oldest known freshwater ray-finned fish (Lackmann et al. 2019). The substantial evidence of cultural transmission in animal populations has provided insight into what should be conserved beyond the protection of genetic diversity and why some conservation programs succeed (by encouraging cultural diversity) while others fail (by neglecting socially transmitted knowledge) (→ Cultural Transmission) (Brakes et al. 2019).

In recognition of the role of animal culture in species' response to anthropogenic disturbance and the enhancement of conservation efforts, a focus on animal culture has been integrated into the conservation initiatives of The Convention on the Conservation of Migratory Species of Wild Animals (CMS), operating under the United Nations Environment Program (UNEP) (Garland et al. 2025). In addition, the open-access Animal Culture Database housed at the University of Arizona currently contains descriptions of 147 behaviors (including vocal communication, migration, predator defense, foraging practices, habitat alteration, play, mating displays, and other social behaviors) for a sample of 66 species across six continents (Basava et al. 2025).

In the words of Brakes et al. (2019), "animal cultures matter for conservation." Social learning can bring about subpopulations with distinctive behaviors that benefit adaptability to change and play a role in establishing and maintaining species' evolutionary trajectories (Brakes et al. 2019). Cultural behaviors contribute to the solidarity and resilience of animal societies, enhancing their ability to thrive in diverse environments and their adaptation to environmental change (→ Adaptation) (→ Global Environmental Change) (Kalan and Luncz 2025).

References

Allen, Jenny A., Ellen C. Garland, Claire Garrigue, Rebecca A. Dunlop, and Michael J. Noad. 2022. "Song Complexity Is Maintained during Inter-Population Cultural Transmission of Humpback Whale Songs." *Scientific Reports* 12: 8999. https://doi.org/10.1038/s41598-022-12784-3.

Appleby, Daniel, Naomi E. Langmore, Ben Pitcher, et al. 2026. "Rescue of the Traditional Song Culture of a Critically Endangered Songbird." *Scientific Reports* 16: 11058. https://doi.org/10.1038/ s41598-026-40115-3.

Basava, Kiran, Md Nafis Ul Alam, Liam Roberts, et al. 2025. "Mapping Nonhuman Cultures with the Animal Culture Database." *Scientific Data* 12 (1019): 1–11. https://doi.org/10.1038/s41597-025-05315-y.

Brakes, Philippa, Sasha R.X. Dall, Lucy M. Aplin, et al. 2019. "Animal Cultures Matter for Conservation." *Science* 363 (6431): 1032–1034.

Brown, Culum, and Michael Webster. 2025. "Fishy Culture in a Changing World." *Philosophical Transactions of the Royal Society B* 380: 20240130. https://doi.org/10.1098/rstb.2024.0130.

Caro, Tim, and Paul W. Sherman. 2012. "Vanishing Behaviors." *Conservation Letters* 5: 159–166.

Crates, Ross, Daniel Appleby, William Bray, Naomi E. Langmore, and Robert Heinsohn. 2025. "Conserving Avian Vocal Culture." *Philosophical Transactions of the Royal Society B* 380: 20240139. https://doi.org/10.1098/rstb.2024.0139.

Garland, Ellen C., Peter Corkeron, Michael J. Noad, et al. 2025. "Culture and Conservation in Baleen Whales." *Philosophical Transactions of the Royal Society B* 380: 20240133. https://doi.org/10.1098/rstb.2024.0133.

Hersh, Taylor A., Daiane S. Marcondes, Gabriel F. Fonseca, et al. 2025. "Ecology and Conservation of Socially Learned Foraging Tactics in Odontocetes." *Philosophical Transactions of the Royal Society B* 380: 20240134. https://doi.org/10.1098/rstb.2024.0134.

IUCN. 2025. International Union for Conservation of Nature Red List of Threatened Species Update, October 10, 2025. www.iucnredlist.org.

Jesmer, Brett, Janey Fugate, and Matthew Kauffman. 2025. "On the Interface between Cultural Transmission, Phenotypic Diversity, Demography and the Conservation of Migratory Ungulates." *Philosophical Transactions of the Royal Society B* 380: 20240131. https://doi.org/10.1098/rstb.2024.0131.

Kalan, Ammie K., and Lydia V. Luncz. 2025. "Saving the Cultural Legacy of Wild Animals." *Science* 388 (6742): 26–27.

Lackmann, Alec R., Allen H. Andrews, Malcolm G. Butler, Ewelina S. Bielak-Lackmann, and Mark E. Clark. 2019. "Bigmouth Buffalo *Ictiobus cyprinellus* Sets Freshwater Teleost Record as Improved Age Analysis Reveals Centenarian Longevity." *Communications Biology* 2 (197): 1–14. https://doi.org/10.1038/s42003-019-0452-0.

Laland, Kevin N., and Vincent M. Janik. 2006. "The Animal Cultures Debate." *Trends in Ecology and Evolution* 21 (10): 542–547.

Roopnarine, Peter D., Roxanne M.W. Banker, and Scott D. Sampson. 2022. "Impact of the Extinct Megaherbivore Steller's Sea Cow (*Hydrodamalis gigas*) on Kelp Forest Resilience." *Frontiers in Ecology and the Environment* 10: 983558. https://doi.org/10.3389/fevo.2022.983558.

Animal Damage Control Act

The Animal Damage Control Act of 1931 (7 U.S.C. §§ 426–426c) is a US federal law that gives the Secretary of Agriculture authority to undertake actions to limit the damage done by wildlife (→ Wilderness Animals) (Pemberton 2020). USDA calls these activities "Wildlife Services" (Public Law 106–387) that are intended to "protect human health and safety, such as in airports and urban areas; manage invasive species, such as feral swine, European starlings, and brown tree snakes (→ Invasive Animal Species); protect agricultural and natural resources, such as livestock (→ Livestock Production), crops, and threatened and endangered species; protect property (→ Property); and monitor wildlife diseases that impact people, domestic pets, livestock, and wildlife" (→ Zoonosis) (US Department of Agriculture Animal and Plant Health Inspection Service 2026). The services include efforts to control bats, coyotes, and raccoons who carry rabies, cervids who carry chronic wasting disease, and migratory birds who prey on aquaculture and agriculture. There is also a program to reduce risk from collisions with animals at airports (→ Conflict with Terrestrial Animals). Approximately 38 percent of the threats are to non-living property, 35 percent to agriculture, 24 percent to health and safety, and 5 percent to natural resources, including endangered species (→ Endangered Species Act).

For over 50 years, much of the effort and considerable controversy has centered around the use of poisons, lethal traps, and aerial hunting to kill large numbers of predators who are considered threats to domesticated cattle regardless of their actual impact. USDA Wildlife Services estimates that the agency kills about two million animals per year, the majority of whom are birds (US Department of Agriculture Wildlife Services, n.d.). In 2024 the killed animals included 354 bears, 23,691 beavers, 57,999 coyotes, 11,227 wild white-tailed deer, 2079 foxes of various species, 12,385 bullfrogs, 8459 raccoons, 246,440 starlings, 127,429 wild swine, 16,667 vultures, and 317 wolves. Wildlife Services also reported 18.5 million animals dispersed to places where they would not do harm to human interests (→ Anthropocentrism).

Critics of this large-scale killing argue that the killing is inhumane, often has adverse effects on non-target species, and is frequently ineffective in achieving the goals of the Animal Damage Control Act (Wood 2015). The criticisms of the Wildlife Services program are part of a larger reconsideration of how to conceptualize and develop approaches to managing human-wildlife conflict (→ Conflict with Terrestrial Animals) (Lamb et al. 2020; Treves and Santiago-Ávila 2020; Whitley et al. 2023). Trying to balance human needs and the impact on other species is a central issue in sustainability. In response to these criticisms, the program has placed greater emphasis on non-lethal alternatives to killing animals, such as diversionary food caches, fladry (placement of flags on fences), livestock protection dogs, and motion-activated lights. The program has reduced the number of animals killed by about 25 percent over the last decade (US Department of Agriculture Animal and Plant Health Inspection Service 2025).

References

Lamb, Clayton T., Adam T. Ford, Bruce N. McLellan, et al. 2020. "The Ecology of Human–Carnivore Coexistence." *Proceedings of the National Academy of Sciences* 117 (30): 17876–17883. https://doi.org/10.1073/pnas.1922097117.

Pemberton, Rachael. 2020. "Detailed Discussion of Wildlife Services." Animal Legal & Historical Center, Michigan State University College of Law. www.animallaw.info/article/detailed-discussion-wildlife-services.

Treves, Adrian, and Francisco J. Santiago-Ávila. 2020. "Myths and Assumptions about Human-Wildlife Conflict and Coexistence." *Conservation Biology* 34 (4): 811–818. https://doi.org/10.1111/cobi.13472.

US Department of Agriculture Animal and Plant Health Inspection Service. 2025. "Wildlife Services Nonlethal Initiative." www.aphis.usda.gov/wildlife-services/about/nonlethal. Accessed 29 December 2025

US Department of Agriculture Animal and Plant Health Inspection Service. 2026. "Wildlife Services Methods Risk Assessments." www.aphis.usda.gov/wildlife-services/about/nonlethal. Accessed 24 June 2026.

US Department of Agriculture Wildlife Services. n.d. "Program Data Reports | Animal and Plant Health Inspection Service." www.aphis.usda.gov/wildlife-services/publications/pdr. Accessed November 19, 2025.

Whitley, Cameron T., Melanie M. Bowers, and Harriett Grantz. 2023. "Urban Human–Coyote Conflicts: Assessing Friendliness as an Indicator of Coexistence." *Animals* 13 (18): 2903.

Wood, Charles. 2015. "Wildlife Disservice: The USDA Wildlife Services' Inefficient and Inhumane Wildlife Damage Management Program." The Humane Society of the United States. www.humaneworld.org/sites/default/files/docs/2015-wildlife-service-white-paper.pdf.

Animal Forests

Animal forests are common and abundant living marine communities dominated by megabenthic suspension feeders (sponges, corals, gorgonians, and bivalves) who form structure, food, and shelter for other sessile and mobile organisms (→ Foundation Species). They are also known as Marine Animal Forests or MAFs. Animal forests are like terrestrial forests but contain animals instead of plants, such as tropical coral reefs (the "rainforests of the sea") and cold-water coral communities. They are one of the most widely distributed ecosystems on the planet because of the wide array of environments that they occupy, from shallow mussel beds to tropical deep cold-water coral communities and Antarctic sponge grounds (Rossi et al. 2017).

Marine animal forests can be dominated by a single species (mussel beds) or by several species (tropical coral reefs). As ecosystem engineers (→ Ecosystem Engineers), the trophic interactions and heterotrophic feeding of marine animal forest suspension feeders are important for nutrient cycling, energy flow within the food web, carbon sequestration, and biodiversity (Denis et al. 2024). Animal forests are hotspots of biodiversity (→ Biodiversity) and play a critical role in the functionality of the oceans. Ecosystems are interdependent so the health of coral reefs is directly related to the health of mangroves, with damage to one impacting the other (Rossi et al. 2017). In addition, coral reef predators such as barracudas help maintain the balance of the reef system by preying on parrotfish who overgraze on algae critical to the benthic ecosystem (→ Trophic Cascades).

Animal forests have been dramatically reduced in the last two decades because of human-induced disturbances to marine ecosystems. Deep-sea species in animal forests have life history characteristics, such as long lives, slow growth, and low reproductive output, that restrict their capacity to recover from human disturbance (Orejas et al. 2022). These animals are impacted by fishing activities, particularly bottom trawling, and shallow coral reefs are disrupted because of their proximity to the shore and human dependency on their resources (→ Ecosystem Services) (Orejas et al. 2022). Shallow water reefs are also impacted by global warming, illegal dumping, recreational fishing, and open-water aquaculture (→ Global Environmental Change) (→ Aquaculture). Coastal tourism is also directly related to the quality of marine ecosystems. Tourism generates income for local human populations, but it also impacts animal forests (→ Wildlife Tourism). For example, scuba diving tourism damages benthic species through unintentional contact, and the probability of contact varies by diver behaviors including the use of limbs, sudden movements, and awareness of the environmental consequences of their actions (Di Franco et al. 2009).

International and national conservation and management initiatives protect some marine animal forests, such as shallow water coral reefs, cold-water coral reefs, and gardens and sponge aggregations, but not as a group together with other ecosystems that have similar ecological roles (Orejas et al. 2022). Effective management and conservation measures for the sustainability of animal forests require a holistic approach that encourages interdisciplinary research and the pooling of different ecosystems under the same umbrella to gain a better understanding of their ecological role (→ Ecocentrism) (Rossi et al. 2017).

References

Denis, Vianney, Christine Ferrier-Pagès, Nadine Schubert, et al. 2024. "Heterotrophy in Marine Animal Forests in an Era of Climate Change." *Biological Reviews* 99: 965–978.

Di Franco, Antonio, Marco Milazzo, Pasquale Baiata, Agostino Tomasello, and Renato Chemello. 2009. "Scuba Diver Behaviour and Its Effects on the Biota of a Mediterranean Marine Protected Area." *Environmental Conservation* 36 (1): 32–40.

Orejas, Covadonga, Marina Carreiro-Silva, Christian Mohn, et al. 2022. "Marine Animal Forests of the World: Definition and Characteristics." *Research Ideas and Outcomes* 8: e96274. https://doi.org/10.3897/rio.8.e96274.

Rossi, Sergio, Lorenzo Bramanti, Andrea Gori, and Covadonga Orejas. 2017. "An Overview of the Animal Forests of the World." In *Marine Animal Forests: The Ecology of Benthic Biodiversity Hotspots*, edited by Sergio Rossi, Lorenzo Bramanti, Andrea Gori, and Covadonga Orejas, 1–28. Springer Nature. https://doi.org/10.1007/978-3-319-21012-4.

Animal-Inclusive Design

Animal-inclusive design (AID) (also known as biodiversity-inclusive design, wildlife-inclusive design, or animal-aided design) is a growing approach in architecture, landscape architecture, and planning that intentionally incorporates the needs of non-human animals into the design of built environments to support biodiversity, ecological resilience, and the broader goals of sustainability. Rather than treating non-human animals as external to human infrastructure, this approach seeks to integrate animal habitats, movement corridors, and the needs of wildlife directly into planning processes (→ Wilderness Animals). The goal is to create spaces where multiple species can not only exist but thrive within urban landscapes (→ Coexistence). This shift coincides with a broader recognition of cities as multispecies environments in which biodiversity and ecosystems are essential components to urban well-being and long-term sustainability (→ Ecosystem Services) (→ Multispecies Futures) (→ Welfare) (Apfelbeck et al. 2020; Hernandez-Santin et al. 2022).

There are numerous projects around the world that embody the principles of animal-inclusive design. Some of the most iconic are wildlife corridors, crossings, or overpasses that connect fragmented spaces to address habitat loss (→ Ecological Connectivity) (→ Habitat Loss) (→ Road Ecology). For instance, in the early 2000s the I-90 Wildlife Bridges Coalition was created to reconnect Washington state's north and south Cascade Mountain regions by establishing safe wildlife crossings across the I-90 highway. An initial bridge was completed in 2018 becoming the largest wildlife overcrossing in all of North America (Conservation Northwest 2021). The Wallis Annenberg Wildlife Crossing has since surpassed this with the I-90 bridge spanning US Highway 101 at Liberty Canyon in California, scheduled for completion in late 2026. The project is a vegetated wildlife overpass measuring approximately 210 feet in length that connects habitat between the Santa Monica Mountains and Simi Hills. The Wallis Annenberg Wildlife Crossing will restore a critical wildlife corridor that has long been fragmented by one of the region's busiest freeways. Because of its scale and design, the structure is widely described as the largest wildlife bridge of its kind in the world (Santa Monica Mountains Conservancy 2026).

The development of large wildlife overpasses and corridors like the one over I-90 and the Wallis Annenberg Wildlife Crossing typically require broad cross-sector collaboration. In many cases, support emerges from a diverse group of actors, including community members, conservationists, hunters, and insurance companies, who come together around a shared objective, even when their motivations or underlying values differ (→ Values). Animal overpasses and corridors restore ecosystem connectivity, enable animals to safely move between fragmented habitats, reduce wildlife-vehicle collisions, and support wildlife and ecosystems in adapting to climate change (→ Ecological Connectivity) (→ Habitat Loss) (→ Road Ecology) (→ Adaptation) (Littlefield et al. 2024; Sugiarto 2023). Comparable infrastructure initiatives now exist in countries across the globe (Orta-Ortiz et al. 2026). Not all efforts require the massive scale of the highway overpass. A coalition in Davis, California, has created enhanced frog habitats by planting native vegetation and placing "toad tunnels" to pass under busy roads (Stewart and Bond 2016). Many urban areas are encouraging plantings that are biodiversity-friendly, drawing on evolving science regarding what is effective (Biella et al. 2025). Another example can be seen in the growing number of buildings and urban parks designed with integrated nesting cavities and habitat features for birds, bats, and other pollinators (→ Liminal Animals) (→ Wilderness Animals). Implementation of animal-inclusive design often involves collaboration between architects and ethologists to identify structural and landscape elements capable of supporting the full life cycles of different species (Apfelbeck et al. 2020; Weisser and Hauck 2025).

Animal-inclusive design is increasingly recognized as a tool for advancing sustainability within urban environments. Sustainable development requires maintaining biodiversity and ecosystem services while meeting human needs, yet urbanization often fragments habitats and reduces species diversity. By embedding habitat features within infrastructure and urban landscapes, animal-inclusive design can help mitigate these effects. Increased biodiversity in cities contributes to ecosystem functions such as pollination, pest regulation, and climate resilience, while also creating opportunities for everyday encounters with nature that can strengthen public support for conservation. Scholars therefore argue that incorporating wildlife into urban planning can produce both ecological and social benefits, linking biodiversity conservation with broader sustainability goals. In this sense, animal-inclusive design represents a practical framework for reimagining cities as multispecies systems, spaces where human well-being, ecological health, and conservation outcomes are mutually reinforcing rather than competing priorities (→ Multispecies Futures) (Baldwin et al. 2011; Visintin et al. 2025).

References

Apfelbeck, Beate, Robbert P.H. Snep, Thomas E. Hauck, et al. 2020. "Designing Wildlife-Inclusive Cities that Support Human-Animal Co-Existence." *Landscape and Urban Planning* 200: 103817.

Baldwin, Robert F., Robert B. Powell, and Stephen R. Kellert. 2011. "Habitat as Architecture: Integrating Conservation Planning and Human Health." *Ambio* 40 (3): 322–327.

Biella, Paolo, Luciano Bani, Enrico Caprio, et al. 2025. "Biodiversity-Friendly Practices to Support Urban Nature across Ecosystem Levels in Green Areas at Different Scales." *Urban Forestry & Urban Greening* 105 (March): 128682. https://doi.org/10.1016/j.ufug.2025.128682.

Conservation Northwest. 2021. "I-90 Wildlife Corridor Campaign: Connecting Wildlife Habitat in the Cascades." Conservation Northwest: Keeping the Northwest Wild. Accessed March 22, 2026. https://conservationnw.org/our-work/habitat/i-90/.

Hernandez-Santin, Cristina, Marco Amati, Sarah Bekessy, and Cheryl Desha. 2022. "A Review of Existing Ecological Design Frameworks Enabling Biodiversity Inclusive Design." *Urban Science* 6 (4): 95.

Littlefield, Caitlin E., Justin P. Suraci, Julia Kintsch, et al. 2024. "Evaluating and Elevating the Role of Wildlife Road Crossings in Climate Adaptation." *Frontiers in Ecology and the Environment* 23 (1): e2816. https://doi.org/10.1002/fee.2816.

Orta-Ortiz, M. Susana, Enzo Falco, Jorge Batista e Silva, et al. 2026. "Are Spatial Plans Transformative for Biodiversity and Ecosystem Services? Insights from Seven European Countries." *European Planning Studies* 34 (2): 235–257. https://doi.org/10.1080/09654313.2025.2582608.

Santa Monica Mountains Conservancy. 2026. "The Wallis Annenberg Wildlife Crossing at Liberty Canyon." Accessed March 21, 2026. https://smmc.ca.gov/liberty-canyon-wildlife-corridor/.

Stewart, Melissa, and Higgins Bond. 2016. *A Place for Frogs*. Revised edition. Peachtree.

Sugiarto, Wisnu. 2023. "Impact of Wildlife Crossing Structures on Wildlife–Vehicle Collisions." *Transportation Research Record* 2677 (2): 670–685.

Visintin, Casey, Georgia E. Garrard, Wolfgang W. Weisser, Mauro Baracco, Richard J. Hobbs, and Sarah A. Bekessy. 2025. "Designing Cities for Everyday Nature." *Conservation Biology* 39 (1): e14328.

Weisser, Wolfgang W., and Thomas E. Hauck. 2025. "Animal-Aided Design–Planning for Biodiversity in the Built Environment by Embedding a Species' Life-Cycle into Landscape Architectural and Urban Design Processes." *Landscape Research* 50 (1): 146–167.

Animal Languages

Research in biology and ethology documents that many non-human animal species have complex and nuanced ways of communicating with members of their own and other species (Meijer 2019a). There are numerous examples of non-human animal communication – dolphins and parrots call each other by name, ravens use referential gestures, the fork- tailed drongo song-bird mimics the alarm calls of other species, and African elephants use different alarm calls to distinguish between different threats such as from bees or from humans (Meijer 2019a). African elephants also address one another with individually specific name-like calls and respond differently to calls directed to them compared to calls directed to another individual (Pardo et al. 2023). In addition, grammatical structures are documented in the communications of many bird species, and prairie dogs are known to "describe humans in detail, including the color of their T-shirts and hair, the speed at which they are approaching, and objects they might be carrying" (Meijer 2019a, 5).

Meijer (2019a) argues that empirical research on animal communication requires us to reconsider the cognitive, social, and linguistic capacities of other animals, including the reconsideration of the very concept of "language" (→ Cognition) (→ Learning for Survival) (→ Animal Cultures). It is important that we ask how humans can interact with other animals to discover their attitudes to the questions that concern their lives. She notes that understanding what animals want involves direct communication and that engaging in dialogue with animals challenges traditional human-animal hierarchies (→ Anthropocentrism) and broadens our concept of language. Animal communication reveals that meaningful expression is not limited to human words, and their forms of communication offer valuable insights into their inner lives and new ways of understanding meaning (→ Multispecies Futures). "For animal languages to be language, other animals do not have to learn anything new; humans just need to begin to see them differently" (Meijer 2019b, 232).

Much of the current research on animal languages uses artificial intelligence to decode animal communication, often based on very large data sets (→ Artificial Intelligence). For example, the Cry Wolf Project, led by Jeffrey T. Reed, has collected the largest database of wild wolf vocalizations in the world, and has now analyzed over 200,000 hours of recordings to decode wolf communication that could transform global conservation efforts (Reed 2025). In addition, the Coller Dolittle Challenge is a competition offering up to $10 million for the use of artificial intelligence that documents two-way interspecies communication. Finally, the Earth Species Project, a nonprofit consortium of researchers, engineers, and ethologists, is devoted to interspecies understanding by using advanced artificial intelligence to unravel animal communication to highlight the voices of nature and rebalance our relationship with the natural world (→ Sustainability) (Wood 2026).

References

Meijer, Eva. 2019a. *When Animals Speak: Toward an Interspecies Democracy*. New York University Press.

Meijer, Eva. 2019b. *Animal Languages*. Translated by Laura Watkinson. The MIT Press.

Pardo, Michael A., Kurt Fristrup, David S. Lolchuragi, et al. 2023. "African Elephants Address One Another with Individually Specific Name-like Calls." *Nature Ecology & Evolution* 8: 1353–1364. https://doi.org/10.1038/s41559-024-02420-w.

Reed, Jeffrey T. 2025. "Can AI Help Us Speak with Wolves?" TED Talk, April 2025. www.thecrywolfproject.com/.

Wood, Drew. 2026. "What Are Animals Saying? AI May Finally Tell Us." A–Z Animals, April 30, 2026. https://a-z-animals.com/articles/what-are-animals-saying-ai-may-finally-tell-us/.

Animal Law

Animal law is the "body of law — statutory and case law — in which the legal, social, or biological nature of non-human animals is an important factor" (Favre 2020, 3). Animal law is ancient – a dozen laws in the Code of Hammurabi (c. 3741–3700 BP) deal with animals (Dedović 2023). In American colonies laws regarding cattle were enacted in the 1600s, and laws about bird hunting in the late 1800s. The field is vast and addresses animal-specific laws that are directly linked to sustainability concerns such as the Animal Damage Control Act, the Convention on International Trade in Endangered Species, the Endangered Species Act, the Marine Mammal Protection Act, and the Pets Evacuation and Transportation Standards Act (→ Animal Damage Control Act) (→ Convention on International Trade in Endangered Species) (→ Endangered Species Act) (→ Marine Mammal Protection Act) (→ Pets Evacuation and Transportation Standards Act). It also includes animal cruelty laws, hunting and game laws, and the treatment and status of animals in legal matters that are not targeted at animals or sustainability, including property and family law. Recently, the field has begun to make comparisons across legal systems and tradition beyond the focus on Western law (Glanert and Mercescu 2024).

Much of the field deals with the theories that determine how animals are defined and treated under laws that are not animal specific. Traditionally in many jurisdictions animals are considered property. Under this legal theory an animal is treated the same way as any inanimate physical object (→ Property). So, for example, if someone harms or kills a companion animal, the human family of the animal is entitled only to the economic value of a "comparable" animal as compensation. This anthropocentric logic is being displaced in many jurisdictions (→ Anthropocentrism). In Ecuador, Bolivia, Colombia, India, New Zealand, and Spain, "nature" and ecosystems have been granted legal standing much like that of humans (→ Ecodemocracy) (Arreaga-Vicuña et al. 2026; Zartner et al. 2025). There are also efforts to provide legal standing to individual non-human animals (Prall 2025). Fitzgerald has documented a growing tendency for US courts to include companion animals in protection orders addressing domestic violence and argues that this is opening a broader view of animals in the legal system (Fitzgerald 2026). Favre has argued for nonhumans to be treated as "living property" with "equitable self-ownership" (Favre 2021). These legal arguments give non-human animals standing to be represented in the existing legal system with broad implications for sustainability so impacts on non-human animals would be given more extensive consideration.

References

Arreaga-Vicuña, Carmen, Mara José Blum-Moarry, David Hoyos-Valarezo, et al. 2026. "Rights of Nature in Global Perspective: Legal Pathways and Policy Implications."

Frontiers in Environmental Science 14 (March): 1647205. https://doi.org/10.3389/fenvs.2026.1647205.
Dedović, Boban. 2023. "Value of Animals – eHammurabi." OMNIKA Foundation. https://ehlaw.org/lch/law-groups/value-of-animals.
Favre, David S. 2020. *Animal Law: Welfare, Interests, and Rights*. 3rd ed. Aspen Coursebook Series. Wolters Kluwer.
Favre, David S. 2021. *The Future of Animal Law*. Edward Elgar Publishing Limited.
Fitzgerald, Amy J. 2026. "Animal-Inclusive Protection Order Statutes in Case Law: Social Problems and the Erosion of the Property Status of Animals." *Social Problems*, February 18, spag012. https://doi.org/10.1093/socpro/spag012.
Glanert, Simone, and Alexandra Mercescu. 2024. "Comparative Animal Law." *The Global Journal of Animal Law* 12 (2): 1–27. https://journal.fi/gjal/article/view/148962/95055.
Prall, Ethan. 2025. "Animal Rights before Legal Personhood." *Cornell Law Review* 110: 75–134.
Zartner, Dana, Fabián A. Cárdenas Castañeda, and Mohammad Golam Sarwar. 2025. *Standing for Nature: Legal Strategies for Environmental Justice*. Island Press.

Animal Research

Animal research refers to the use of non-human animals to investigate biological processes, develop medical treatments, test product safety, or understand behavior and cognition. These studies can occur in laboratory settings, agriculture facilities, or controlled field environments, and they typically involve regulated protocols designed to minimize pain and distress. In the US whether animals used in research are afforded legal protections or rights depends on their species designation and regulatory classification (→ Rights). As a result, rats, mice, and birds are bred for research but are excluded from basic welfare (→ Welfare) standards and rights frameworks applied to other sentient animals (→ Sentience).

Proponents of animal research argue that it has contributed significantly to biomedical advances, while critics emphasize the ethical implications of using sentient beings as experimental subjects. Abolitionists completely reject the idea of using animals in research regardless of the context (→ Abolitionism). Contemporary scholarship highlights that animal research is shaped by cultural norms, institutional priorities, and ethical frameworks. For example, the "Three Rs" (replacement, reduction, and refinement) seek to limit harm while maintaining animal use in research. Scholars argue that these aspects need to constantly be assessed and reformed (Smith and Lilley 2019; DeGrazia and Beauchamp 2019).

An example of the complexity of animal research can be found in the use of rodents for biomedical studies. Mice and rats are commonly used to model human disease because their genetic, physiological, and behavioral traits can be studied in controlled conditions. However, these animals also exhibit sophisticated cognitive and emotional capacities, including empathy-oriented behaviors and social bonding (→ Cognition) (→ Empathy) raising important questions about their welfare (→ Welfare) (Mason 2021; Bartal et al. 2011). In addition, studies have shown that housing social animals in laboratory environments, such as in barren cages or with restricted access to conspecifics, can significantly alter rodents' stress response and behavior, potentially affecting both their well-being and the reliability of scientific data (Massari et al. 2018). This interplay between scientific goals and animal welfare demonstrates why animal research remains ethically and scientifically debated.

Laboratory and agriculture research facilities consume energy, water, and materials, generating environmental footprints that must be considered in broader sustainability planning. In some cases, research on laboratory animals is done to better understand the impacts of pollution and other environmental stressors and research on animals in zoos and similar settings is used to support the protection of threatened species and to restore ecosystems affected by climate change, pollution, and habitat loss (→ Restoration Ecology) (→ Habitat Loss) (→ Pollution) (Malhi et al. 2020). Emerging alternatives, such as in vitro models, computational simulations, artificial intelligence, and organ-on-a-chip technologies, are increasingly promoted not only for ethical reasons but also because they reduce resource use and waste. By assessing the ecological costs and benefits of research practices, sustainability frameworks encourage more responsible, transparent, and innovative agendas.

References

Bartal, Inbal Ben-Ami, Jean Decety, and Peggy Mason. 2011. "Empathy and Pro-Social Behavior in Rats." *Science* 334 (6061): 1427–1430.

DeGrazia, David, and Tom L. Beauchamp. 2019. "Beyond the 3 Rs to a More Comprehensive Framework of Principles for Animal Research Ethics." *ILAR Journal* 60 (3): 308–317.

Malhi, Yadvinder, Janet Franklin, Nathalie Seddon, et al. 2020. "Climate Change and Ecosystems: Threats, Opportunities and Solutions." *Philosophical Transactions of the Royal Society B* 375 (1794): 20190104.

Mason, Peggy. 2021. "Lessons from Helping Behavior in Rats." *Current Opinion in Neurobiology* 68: 52–56.

Massari, Catia Helena de Almeida Lima, Nadini Oliveira Martins, Angela Faustino Jozala, Denise Grotto, and Marli Gerenutti. 2018. "Laboratory Animal Welfare: Environmental Enrichment Shows Positive Effect on Animal Testing." *Brazilian Journal of Veterinary Research and Animal Science* 55 (4): e145008.

Smith, Adrian J. and Elliot Lilley. 2019. "The Role of the Three Rs in Improving the Planning and Reproducibility of Animal Experiments." *Animals* 9 (11): 975.

Animal Welfare Act

The Animal Welfare Act (AWA) (7 U.S.C. §§ 2131–2159) is the only federal law regulating the treatment of animals used in research, teaching, transportation, and exhibition in the United States. The AWA was first signed into law in 1966 as the Laboratory Animal Welfare Act (P.L. 89–544) (→ Animal Research). The passage of the initial AWA was motivated by public concern over the use of species that are often companion animals in biomedical research and the conditions under which such animals were housed and transported (→ Domestic Animals). Six species were covered by the 1966 AWA: dogs, cats, non-human primates, guinea pigs, hamsters, and rabbits. The Act established a baseline of oversight for animal welfare by requiring animal dealers, research facilities, and exhibitors to register with the US Department of Agriculture (USDA) and comply with standards regulating adequate housing, veterinary care, sanitation, and handling (→ Welfare). The AWA has been amended eight times (in 1970, 1976, 1985, 1990, 2002, 2007, 2008, and 2013). The AWA has never applied to invertebrates (Metzger 2015).

The 1970 amendment expanded the AWA to protect all warm-blooded animals (excluding farm animals) and extended registration to animal exhibitors like carnivals, circuses, zoos, and aquariums (→ Wilderness Animals) (→ Livestock Production) (→ Zoos and Aquariums) (Lee 2016). In 1976, an amendment expanded transportation coverage to include airline, railroad, motor carrier, and shipping lines (Lee 2016). The 1985 amendments improved standards for laboratory animals by requiring enrichment, providing exercise, considering alternatives to painful and/or distressing procedures, and establishing the Institutional Animal Care and Use Committees (IACUCs) (Lee 2016). In 1990, the AWA was amended to require that all dogs and cats be held at shelters for a minimum of five days before being sold to a research facility to give time for recovery or adoption. Twelve years later, the AWA was amended to limit the definition of "animal" to exclude birds, rats, and mice bred for research (Lee 2016). Then, in 2007, amendments prohibited the attachment of knives and other items on birds for fighting purposes, while also increasing penalties for animal fighting (Lee 2016). Further protections for dogs were included in 2008 amendments, stipulating that dogs imported to the US be vaccinated, in good health, and at least six months of age (Lee 2016). Dog fighting prohibitions were also strengthened and fines for violations were increased. In 2013, the definition of "exhibitor" was changed to make a distinction between small and large-scale exhibitors, so that people exhibiting personal animals who reside in the home generating limited income would be excluded from licensing requirements (Lee 2016).

Today, the AWA regulates a range of settings where animals are used, including biomedical laboratories, zoos, circuses, breeding facilities, and commercial transport operations (→ Animal Research) (→ Zoos and Aquariums) (→ Wildlife Trade). For example, research institutions that use animals such as dogs or primates must maintain records, provide veterinary care, and subject research protocols to be reviewed by an IACUC. Similarly, animal exhibitors like zoos, aquariums, and circuses, must meet federal standards regarding enclosure size, sanitation, and animal care.

Despite these protections, the AWA still has notable limitations. The definition of "animal" excludes farm animals used for food or fiber production, cold-blooded species such as reptiles and fish, most rats, mice, and birds bred for research, and all invertebrates (→ Livestock Production) (→ Animal Research) (McFadden et al. 2019). The AWA's exclusion of cephalopods (octopus, squids, cuttlefish and nautiluses) and decapod crustaceans (crabs, lobsters, shrimp, and prawns) has been challenged (National Institutes of Health 2023), due to numerous studies highlighting their sentience and intelligence (→ Sentience) (→ Cognition) (Schnell et al. 2021; Marino 2017; Mather and Dickel 2017). One component of the AWA is the regulation of the Institutional Animal Care and Use Committee (IACUC), which is a federally mandated committee that oversees animal research at institutions to ensure that welfare and ethical standards are maintained. Although invertebrates are not protected directly under the AWA, many research institutions now require IACUC review for cephalopod research (Leland and Gelperin 2025).

Although the Animal Welfare Act was not originally intended as environmental legislation, nor specifically designed to protect wild or wilderness animals, it nonetheless carries important implications for sustainability (→ Wilderness Animals). In practice, the AWA has been used to regulate the treatment of certain wild mammalian species such as tigers, bears, and primates who are held in captivity in research facilities, exhibitions, or in other managed environments. Broadly, animal welfare regulation shapes how humans view and interact with animals in research, education, and entertainment, thereby influencing broader ethical norms about multispecies coexistence (→ Coexistence). By setting baseline standards for humane care, the AWA encourages institutions to consider the well-being of animals alongside scientific and economic goals. However, there are many critiques. Some argue that the AWA's exclusions highlight ongoing debates between welfare regulation, science, and economic goals. Some scholars argue that the AWA is ineffective and underenforced (Vanderau 2006; Rudloff 2017) and that strengthening and modernizing the AWA by expanding species coverage and improving enforcement mechanisms could help align animal welfare policy with emerging sustainability frameworks that emphasize ethical responsibility toward non-human life (→ Biocentrism).

References

Lee, Courtney G. 2016. "The Animal Welfare Act at Fifty: Problems and Possibilities in Animal Testing Regulation." *Nebraska Law Review* 95: 194.

Leland, Stuart, and Alan Gelperin. 2025. "IACUC Considerations for Research Using Invertebrates." In *The IACUC Handbook*, edited by Mark Suckow, Sreekant Murthy, Julie M. Sharp, and Troy Hallman. CRC Press. https://doi.org/10.1201/9781003272861.

Marino, Lori. 2017. "Cetacean Cognition." In *The Oxford Handbook of Animal Studies*, edited by Linda Kalof. Oxford University Press.

Mather, Jennifer A., and Ludovic Dickel. 2017. "Cephalopod Complex Cognition." *Current Opinion in Behavioral Sciences* 16: 131–137.

McFadden, Michael, Sue Leary, and Kathy Hessler. 2019. "Animal Welfare Act: Excluded Animals." *Animal Law Review* 25 (2): 203–223.

Metzger, Mitchell M. 2015. "Knowledge of the Animal Welfare Act and Animal Welfare Regulations Influences Attitudes toward Animal Research." *Journal of the American Association for Laboratory Animal Science* 54 (1): 70–75.

National Institutes of Health. 2023. "Request for Information (RFI) on Proposed Guidance to Assured Institutions on Cephalopod Care and Use" (Notice number NOT-OD-23-176). National Institutes of Health, Office of Laboratory Animal Welfare. https://grants.nih.gov/grants/guide/notice-files/NOT-OD-23-176.html.

Rudloff, Leslie. 2017. "Failure to Launch: The Lack of Implementation and Enforcement of the Animal Welfare Act." *Syracuse Law Review* 67: 173–190.

Schnell, Alexandra K., Piero Amodio, Markus Boeckle, and Nicola S. Clayton. 2021. "How Intelligent Is a Cephalopod? Lessons from Comparative Cognition." *Biological Reviews* 96 (1): 162–178.

Vanderau, Melanie L. 2006. "Science at Any Cost: The Ineffectiveness and Underenforcement of the Animal Welfare Act." *Penn State Environmental Law Review* 14 (3): 721–742.

Anthropocene

It is argued that the Holocene geological epoch has ended and that the Earth had entered a new epoch, the Anthropocene (Crutzen and Stoermer 2000). The Holocene began with the end of widespread glaciations ("ice ages") about 12,000 years ago and is characterized by relatively stable global temperatures, atmospheric chemistry, and sea level. Crutzen and Stoermer (2000) argue that human impacts on Earth had become so substantial that the changes would be detectable in the geological record in the far distant future (→ Global Environmental Change). They propose that the Anthropocene started when the Industrial Revolution at the end of the eighteenth century led to the massive increase in fossil fuel use and thus of greenhouse gases and other human-generated compounds in the atmosphere.

The term has become popular as a way of signaling the massive impacts humans have on the biosphere. Several alternative starting points for the Anthropocene have been proposed. Some suggest that the rise of agriculture about 10,000 years ago or the massive changes that came with the "Colombian Exchange" and the incursion of Europeans into the Americas with the accompanying ecological and social changes in the sixteenth century would be appropriate marking points of the Anthropocene. However, others argue that the Anthropocene began with the "Great Acceleration" that occurred after 1945, a period marked by rapid growth in the population of humans and domesticated animals (→ Domestic Animals), introduction to the environment of novel compounds such as plastics and radioactive isotopes, and substantial changes in the chemistry of Earth's atmosphere (Zalasiewicz et al. 2024).

The concept of the Anthropocene has not been formally accepted as a designated geological epoch for technical reasons (Witze 2024), but there is wide acceptance of the concept even without a formal geological designation (Edgeworth et al. 2024). For example, the Anthropocene links closely to analyses of planetary boundaries (→ Planetary Limits) that indicate when human actions are generating global changes that degrade hopes for sustainability.

References

Crutzen, Paul J., and Eugene F. Stoermer. 2000. "The Anthropocene." *IGBP Newsletter*, no. 41: 17–18. www.igbp.net/download/18.316f18321323470177580001401/1376383088452/NL41.pdf.

Edgeworth, Matthew, Andrew M. Bauer, Erle C. Ellis, et al. 2024. "The Anthropocene Is More than a Time Interval." *Earth's Future* 12 (7): e2024EF004831. https://doi.org/10.1029/2024EF004831.

Witze, Alexandra. 2024. "Geologists Reject the Anthropocene as Earth's New Epoch — after 15 Years of Debate." *Nature* 627(March): 249–250. https://doi.org/10.1038/d41586-024-00675-8.

Zalasiewicz, Jan, Julia Adeney Thomas, Colin N. Waters, Simon Turner, and Martin J. Head. 2024. "What Should the Anthropocene Mean?" *Nature* 632 (August): 980–984.

Anthropocentrism

Anthropocentrism is the belief that only humans have intrinsic value, and all other beings have value only in their ability to serve humans (their instrumental value) (Goralnik and Nelson 2012). In this view, humans possess direct moral standing because they are ends in and of themselves, and other living beings and systems are means to human ends (Goralnik and Nelson 2012). There are two types of anthropocentrism: 1) narrow or strong (the view of humans in isolation and relationships with non-humans is unimportant) and 2) broad or weak (the view that humans exist in ecological contexts, embedded in and dependent on relationships with other beings and systems, although non-human nature is still valued only in its relation to humans) (→ Ecosystem Services) (Goralnik and Nelson 2012,149).

The first prominent environmental philosopher to advocate for narrow anthropocentrism was John Passmore (1974). Passmore argued that the contemporary ecological problems of the Western world cannot be resolved by adopting a *natura naturans* (nature doing what nature does) perspective, but rather by using nature to fulfill human needs – "only by transforming nature can it continue to survive … [transformation] can make the world more fruitful, more diversified, and more beautiful" (Passmore 1974, 179).

Ten years later, Bryan Norton (1984) proposed a "weak anthropocentrism" that is adequate to support an environmental ethic without having to attribute intrinsic value to nature as in ecocentrism (→ Ecocentrism). In weak anthropocentrism, only humans have moral standing, but the well-being of the natural world has instrumental value for what it brings to human welfare, so humans have an obligation to protect the environment and other species. He argues that "weak anthropocentrism provides a framework adequate to criticize current destructive practices to incorporate concepts of human affinity to nature, and to account for the distinctive nature of environmental ethics" (Norton 1984, 148).

Hayward (1997) argues that criticisms of anthropocentrism are better understood as critiques of 1) speciesism (arbitrary discrimination on the basis of species) and 2) human chauvinism (identification of differences in ways that invariably favor humans). In addition, he contends that anthropocentrism is an element of any ethic: "if the ultimate point of an ethic is to yield a determinate guide to human action, then, the human reference is ineliminable even when extending moral concern to nonhumans" (Hayward 1997, 56). Finally, anti-anthropocentrism is counterproductive in practice because of the failure to distinguish between legitimate human interests (the need to care for other members of one's own species) and illegitimate human interests (speciesism and human chauvinism) (Hayward 1997, 59). Thus, anthropocentrism makes ethics possible because it is a basic feature of the logic of obligation: "if an ethic is a guide to action, and if a particular ethic requires an agent to make others' ends her ends, then they become just that – the agent's ends," and this is a limitation in the construction of a completely nonanthropocentric ethic (→ Values) (Hayward 1997, 57).

In a critique of Hayward's above arguments, Kopnina et al. (2018) claim that anthropocentrism is a significant driver of ecocide and the environmental crisis, cannot lead us to sustainability, and ignores the fact that humanity is fully dependent on nature and the web of life (→ Ecocentrism). While debate among ethicists on the meaning and implications of anthropocentrism continues, the term remains useful in flagging perspectives that give consideration to non-humans only to the extent that they are valuable to humans.

References

Goralnik, Lissy, and Michael P. Nelson. 2012. "Anthropocentrism." *Encyclopedia of Applied Ethics*, 2nd ed., edited by Ruth Chadwick, 145–155. Elsevier.

Hayward, Tim. 1997. "Anthropocentrism: A Misunderstood Problem." *Environmental Values* 6 (1): 49–63.

Kopnina, Helen, Haydn Washington, Bron Taylor, and John J. Piccolo. 2018. "Anthropocentrism: More than Just a Misunderstood Problem." *Journal of Agricultural and Environmental Ethics* 31: 109–127.

Norton, Bryan. 1984. "Environmental Ethics and Weak Anthropocentrism." *Environmental Ethics* 6 (2): 131–148.

Passmore, John. 1974. *Man's Responsibility for Nature: Ecological Problems and Western Traditions*. Charles Scribner's Sons.

Applied Zooarchaeology

Applied zooarchaeology is an approach to conservation biology that includes both paleontological and zooarchaeological data to provide temporal depth to information on biological and cultural changes relevant to the current environmental crises (→ Conservation Biology) (Wolverton and Lyman 2012). Paleontological collections of ancient animal remains provide valuable data on natural processes in the absence of human behaviors, and zooarchaeology provides information on the evolutionary and ecological implications of animal-human interactions (Wolverton and Lyman 2012). Some of the questions answered by applied zooarchaeology include the identification of species who are native (→ Native Species) and occurred in an area in the past, what species have been introduced by humans and/or are exotic (→ Invasive Animal Species), and what kinds of changes to ecological communities have been caused by humans (→ Global Environmental Change). Applied zooarchaeology also provides a way to understand the evolutionary trajectories of both terrestrial and aquatic ecosystems. For example, detection of Holocene changes in sea mammal populations on the Northwest coast of the US provides vital information for the conservation efforts there (→ Restoration Ecology). Finally, applied zooarchaeology offers conservation biology a deep time perspective on historically contingent processes, prompting recognition of the need to value all elements in nature in pursuit of well-being and sustainability (→ Ecocentrism).

Zooarchaeology also challenges the foundations of anthropocentrism. Examining animal remains offers valuable insights into biodiversity, such as species introductions and extinctions, which inform sustainable practices and support conversations about re-wilding in the Anthropocene (→ Anthropocentrism) (→ Biodiversity) (→ Rewilding) (→ Anthropocene) (Sykes 2014). For example, scholars have examined the biocultural history of fallow deer – who have been repeatedly relocated, wiped out, and reintroduced over the past 10,000 years – using zooarchaeological, biomolecular, historical, and iconographic evidence (Baker et al. 2024). Their work highlights the difficulty in classifying species as wild or domestic. European fallow deer, despite being farmed and showing signs of selective breeding, are often excluded from both scholarly reviews of domestic animal management and conservation efforts due to their ambiguous domestication history (→ Domestic Animals) (Baker et al. 2024). Applied zooarchaeology not only enriches conservation biology with long-term ecological and cultural perspectives, but also destabilizes rigid human-animal binaries, underscoring the importance of historically informed approaches to biodiversity and sustainability in the Anthropocene.

References

Baker, Karis H., Holly Miller, Sean Doherty, et al. 2024. The 10,000-year Biocultural History of Fallow Deer and Its Implications for Conservation Policy. *Proceedings of the National Academy of* Sciences 121(8): e2310051121.

Sykes, Naomi. 2014. *Beastly Questions: Animal Answers to Archaeological Issues*. Bloomsbury.

Wolverton, Steve, and R. Lee Lyman. 2012. "Introduction to Applied Zooarchaeology." In *Conservation Biology and Applied Zooarchaeology*, edited by Steve Wolverton and R. Lee Lyman. The University of Arizona Press.

Aquaculture

Aquaculture is the farming of fish, shellfish, and aquatic plants in controlled environments. Humans have been practicing aquaculture for thousands of years based on the evidence of tilapia harvesting from ponds in Egypt 4500 years ago (Ahmed and Thompson 2019). The rapid increase in aquaculture as a vital agricultural activity that contributes to human nutrition is called the Blue Revolution, a phenomenon that is occurring in most parts of the world. Before 1980 aquaculture was a relatively insignificant global food source, but by 2015 it had grown to be a larger source of seafood for human consumption than wild fisheries and currently takes place almost exclusively in developing countries (Garlock et al. 2023). Aquaculture accounted for 46 percent of global fish production in 2018 and is a significant addition to a global food system impacted by climate change and the ever-increasing demand for animal protein (Jiang et al. 2022).

There are two types of aquaculture environments: freshwater and marine. Freshwater aquaculture occurs in inland water bodies such as rivers, lakes, and ponds and typically grows freshwater species such as tilapia and catfish. Marine aquaculture cultivates marine species such as salmon and sea bass in marine environments including sea cages and offshore farms. Mariculture is another type of marine farming that takes place exclusively in marine saltwater environments such as seas, oceans, and estuaries and where only high-value marine species are farmed including barramundi, salmon, seabass, seabream, trout, bivalve mollusks (clams, mussels, oysters, scallops), and seaweeds.

Aquaculture farming places substantial demand on water and energy resources, and in most systems, aquaculture requires higher inputs of energy than livestock production (→ Livestock Production) (Jiang et al. 2022). The impact of aquaculture on the environment is also substantial in its contribution to greenhouse gas emissions, eutrophication, and the degradation of aquatic, benthic, and coastal habitats and ecosystems (→ Animal Forests) (Jiang et al. 2022). One strategy to enhance the use of land and water while reducing environmental risks is rice-animal co-culture that combines animal production (such as fish, shellfish, crab, shrimps, eels, frogs, turtles, and ducks) with the production of rice in paddy systems. Rather than using chemicals to control pests, animals eat plant hoppers, weeds, and pests while providing nutrients for the rice; animal activities in the paddies help loosen the soil improving soil permeability and oxygen content; and rice plants provide shade for the animals and help lower the temperature of the standing water (Bashir et al. 2019; Ge et al. 2023). Rice-animal co-culture reduces environmental pollution and greenhouse gas emissions, maintains soil fertility, stabilizes grain incomes, and preserves paddy field biodiversity (→ Biodiversity) (→ Pollution) (Ge et al. 2023).

Both freshwater and marine aquaculture perform similarly in aggregate on three dimensions of sustainability (economic, social, environmental) so the type of water body makes little difference from a sustainability perspective

(Garlock et al. 2023). However, there are variations in sustainability across different aquaculture systems, and the systems in the Global North score higher in all three dimensions of sustainability compared to the developing countries (Garlock et al. 2023). Jiang et al. (2022) developed a food-energy-water-carbon sustainability index (scored 0–100) to assess the sustainability of global aquaculture across countries. They found the overall sustainability of global aquaculture was low (average score = 26), no country achieved a high score (75–100), and almost all practiced aquaculture in a low-sustainable way (0–50). China contributed more than half of global aquaculture water consumption and greenhouse gas emissions, followed by India and Indonesia.

References

Ahmed, Nesar, and Shirley Thompson. 2019. "The Blue Dimensions of Aquaculture: A Global Synthesis." *Science of the Total Environment* (652): 851–861. https://doi.org/10.1016/j.scitotenv.2018.10.163.

Bashir, Muhammad Amjad, Jian Liu, Yucong Geng, et al. 2019. "Co-culture of Rice and Aquatic Animals: An Integrated System to Achieve Production and Environmental Sustainability." *Journal of Cleaner Production* 249: 1–11. https://doi.org/10.1016/j.jclepro.2019.119310.

Garlock, Taryn M., Frank Asche, James L. Anderson, et al. 2023. "Environmental, Economic, and Social Sustainability in Aquaculture: The Aquaculture Performance Indicators." *Nature Communications* 15: 5274. https://doi.org/10.1038/s41467-024-49556-8.

Ge, Lei, Yu Sun, Yujie Li, et al. 2023. "Ecosystem Sustainability of Rice and Aquatic Animal Co-culture Systems and a Synthesis of Its Underlying Mechanisms." *Science of the Total Environment* 880: 163314. http://dx.doi.org/10.1016/j.scitotenv.2023.163314

Jiang, Qutu, Nishan Bhattarai, Markus Pahlow, and Zhenci Xu. 2022. "Environmental Sustainability and Footprints of Global Aquaculture." *Resources, Conservation & Recycling* 180: 106183. https://doi.org/10.1016/j.resconrec.2022.106183.

Artificial Intelligence (AI)

Artificial intelligence (AI) is the ability of machine-based systems to carry out tasks associated with intelligence in humans, including decision making, pattern recognition, problem solving, and reasoning (→ Cognition). The possible benefits and risks of the increasing capabilities and prevalence of AI are frequently debated. For example, more than 1000 AI experts have signed a statement that "Mitigating the risk of extinction from AI should be a global priority alongside other societal-scale risks such as pandemics and nuclear war" (Hinton et al., n.d.). Few of these discussions have considered the implications of AI for non-human animals, but the impacts of AI might be very substantial for them (Foris and Rault 2025; Coghlan and Parker 2023; Singer and Tse 2023). For example, AI is used to manage herds of domesticated animals, but it is not clear that such AI takes account of animal welfare concerns (→ Livestock Production) (→ Welfare). There is also the potential that AI could be deployed to enhance the well-being of non-human animals, for example using AI systems to prevent poaching (→ Wildlife Trade), better understand habitat use, match animals in shelters with potential adopters, reduce road kills via self-driving vehicles, lead to better veterinary treatment of livestock and zoo animals, and decode non-human animal language (→ Road Ecology) (→ Zoos and Aquariums) (→ Animal Languages) (Coghlan and Parker 2024; Bossert 2023, Liptovsky and Polla 2025).

As AI systems increase in capabilities, they may develop features that seem to constitute self-awareness. Self-awareness in non-human animals plays a role in many ethical arguments, raising the issue of whether the same arguments that justify an ethical concern for non-human animals would also call for ethical concern for AIs (→ Cognition) (Coeckelbergh 2020; Müller 2021).

AI also creates risks and potential benefits for overall sustainability that will have implications for non-human animals (Toderas 2025; Berthelot et al. 2025). AI might provide guidance that would aid in achieving sustainability, but it is also developing a very substantial ecological footprint through the excessive amount of energy and water required to support computing (→ Ecological Footprint).

References

Berthelot, Adrien, Eddy Caron, Mathilde Jay, and Laurent Lefèvre. 2025. "Understanding the Environmental Impact of Generative AI Services." *Communications of the ACM* 68 (7): 46–53. https://doi.org/10.1145/3725984.

Bossert, Leonie N. 2023. "Benefitting Nonhuman Animals with AI: Why Going Beyond 'Do No Harm' Is Important." *Philosophy & Technology* 36 (3): 57. https://doi.org/10.1007/s13347-023-00658-z.

Coeckelbergh, Mark. 2020. *AI Ethics*. The MIT Press Essential Knowledge Series. The MIT Press.

Coghlan, Simon, and Christine Parker. 2023. "Harm to Nonhuman Animals from AI: A Systematic Account and Framework." *Philosophy & Technology* 36 (2): 25. https://doi.org/10.1007/s13347-023-00627-6.

Coghlan, Simon, and Christine Parker. 2024. "Helping and Not Harming Animals with AI." *Philosophy & Technology* 37 (1): 20. https://doi.org/10.1007/s13347-024-00712-4.

Foris, Borbala, and Jean-Loup Rault. 2025. "AI-Agent Ethics Should Consider Sentient Non-Human Animals." *Nature* 645 (8079): 41. https://doi.org/10.1038/d41586-025-02796-0.

Hinton, Geoffrey, et al. n.d. "Statement on AI Risk | CAIS." Center for AI Safety. Accessed September 21, 2025. https://aistatement.com.

Liptovsky, Matyan, and Emily Polla. 2025. "Artificial Intelligence's Potential in Zoo Animal Welfare." *Zoo Biology* 44 (5): 403–411.

Müller, Vincent C. 2021. "Is It Time for Robot Rights? Moral Status in Artificial Entities." *Ethics and Information Technology* 23 (4): 579–587. https://doi.org/10.1007/s10676-021-09596-w.

Singer, Peter, and Yip Fai Tse. 2023. "AI Ethics: The Case for Including Animals." *AI and Ethics* 3 (2): 539–551. https://doi.org/10.1007/s43681-022-00187-z.
Toderas, Mihaela. 2025. "Artificial Intelligence for Sustainability: A Systematic Review and Critical Analysis of AI Applications, Challenges, and Future Directions." *Sustainability* 17 (17): 8049. https://doi.org/10.3390/su17178049.

Artificial Selection and Breeding

Artificial selection and breeding refer to the human-directed process of choosing specific animals with desirable traits to reproduce, thereby shaping the genetic characteristics of future generations. Artificial selection differs from natural selection which is driven by environmental pressures and survival advantages (→ Evolution). Over time, selective breeding has produced dramatic changes in domesticated species from the development of specialized livestock breeds to companion animals with distinctive physical and behavioral traits (→ Domestic Animals). The changes humans have induced via breeding are so extensive that Charles Darwin invoked them in support of the changes that might be produced via natural selection (Darwin 1868).

Artificial selection is both a biological and cultural practice, influenced by economic systems, technological advancements, and societal values (→ Values). Understanding this process helps explain the diversity of domestic animals and the ethical questions raised by intensive breeding. Examples of artificial selection and breeding span a wide range of species and illustrate how human priorities shape animal bodies and behaviors. Broiler chickens, for instance, have been selectively bred for accelerated growth rates. Many domestic dog breeds reflect aesthetic rather than functional priorities, with traits that can restrict breathing and shorten lifespans. Laboratory mice are routinely selectively bred for specific genetic or behavioral traits to model human diseases (→ Animal Research). Aquaculture species such as salmon and tilapia are bred to maximize growth efficiency under captive conditions (→ Aquaculture).

In the example of broiler chickens, selective breeding for rapid growth and high meat yield has produced birds who reach market weight in a fraction of the time required by their mid-twentieth-century counterparts. While this increases economic efficiency, it has also created significant welfare concerns (→ Welfare). Broiler chickens often experience skeletal deformities, cardiovascular strain, and limited mobility due to extremely rapid growth (Tainika et al. 2023). Similarly, modern dairy cows have been selectively bred for extremely high milk production, resulting in animals who produce far more milk than their bodies evolved to sustain. While this has increased productivity within the dairy industry, it has also been associated with elevated rates of lameness, mastitis, metabolic disorders, and reduced lifespan. As with broiler chickens, these outcomes illustrate how selective breeding practices often prioritize output and efficiency while externalizing the physical costs borne by the animals themselves.

Artificial selection and breeding have profound implications for sustainability. Livestock bred for high productivity can require intensive resource inputs in terms of feed, water, land, and energy, contributing to environmental pressures such as greenhouse gas emissions and habitat conversion. The waste products produced from intensive livestock production can have environmental justice implications (→ Environmental Justice). At the same time, selective breeding can support sustainability goals when used to enhance disease resistance, improve food availability, or adapt animals to changing climate conditions (Gjedrem et al. 2012; Conner 2003). For instance, some scholars have suggested that selective breeding of insects could better help feed the world's population (→ Insects as Food) (Hansen et al. 2025). However, reliance on genetically uniform breeds can reduce biodiversity and increase vulnerability to disease outbreaks (→ Biodiversity) (→ Zoonosis). Sustainable breeding strategies involve balancing productivity with genetic diversity, ecological resilience, and animal welfare. Recent advances in genetic technology allow the production of animals with desired traits without having to artificially select over multiple generations but can lead to the same problems as breeding for human goals (→ Genetically Modified Organisms).

References

Conner, Jeffrey K. 2003. "Artificial Selection: A Powerful Tool for Ecologists." *Ecology* 84 (7): 1650–1660.

Darwin, Charles. 1868. *The Variation of Animals and Plants Under Domestication.* Volume 2. John Murray.

Gjedrem, Trygve, Nick Robinson, and Morten Rye. 2012. "The Importance of Selective Breeding in Aquaculture to Meet Future Demands for Animal Protein: A Review." *Aquaculture* 350: 117–129.

Hansen, Laura Skrubbeltrang, Stine Frey Laursen, Simon Bahrndorff, et al. 2025. "The Unpaved Road towards Efficient Selective Breeding in Insects for Food and Feed—A Review." *Entomologia Experimentalis et Applicata* 173 (6): 498–521.

Tainika, Brian, Ahmet Şekeroğlu, Ahmet Akyol, and Zacharia Waithaka Ng'ang'a. 2023. "Welfare Issues in Broiler Chickens: Overview." *World's Poultry Science Journal* 79 (2): 285–329.

B

Biocentrism

Biocentrism is the view that all living things have value and moral status, including bacteria, trees, and animals. Some Eastern religions such as Jainism (→ Religion) have long been biocentrist, but the first prominent Western thinker to advocate for biocentrism was Albert Schweitzer who coined the phrase "reverence for life" in 1923 (Attfield 2016). Schweitzer believed that all living individuals have a "will to live" (or a built-in drive toward self-preservation), and the reverence that one feels toward one's own "will to live" ought to be extended to all (Attfield 2016). Biocentrism gives value to non-humans as well as humans but does not extend moral consideration to ecosystems as wholes or to their abiotic components (→ Ecocentrism) (Kopina and Washington 2020). Positions labeled "-centrist" are often associated with an emotional attachment to certain entities, such as biophilia (love of life) (→ Biophilia), zoophilia (love of animals), or arborphilia (love of trees) (Kopina and Washington 2020).

A biocentric, life-centered theory of environmental ethics was first proposed as a "Respect for Nature" by Paul Taylor in 1981. He argues that all life forms have inherent value independent of their utility to humans. He argues that three interrelated components must be adopted to achieve sustainability: 1) a moral attitude of "respect for nature"; 2) a belief system that conceives of the natural world and of our place in it that supports respect for Earth's natural ecosystems and their life communities; and 3) a system of moral rules for guiding our treatment of those ecosystems and life communities (Taylor 1981, 197). His perspective is biocentric (life-centered not sentience-centered), individualistic (individual organisms have moral standing but not species or ecosystems), and egalitarian (all organisms including humans have equal inherent worth).

The adoption of a biocentric outlook as a guide for sustainability promotes biodiversity, human and animal coexistence, the restoration of degraded ecosystems, and informs the establishment of protected areas that emphasize the intrinsic worth of wild spaces and their inhabitants (→ Biodiversity) (→ Coexistence) (→ Rewilding) (→ Ecological Restoration) (→ Ecological Connectivity). Attfield (2011) argues that the best way to face climate change is with a biocentric environmental ethic; while ecosystems are important, their value lies in the value of the creatures

DOI: 10.4324/9781003528142-3

that they support, not in some independent value of their own (→ Climate Change) (→ Ecocentrism).

References

Attfield, Robin. 2011. "Climate Change, Environmental Ethics and Biocentrism." In *Climate Change and Environmental Ethics*, edited by Ved Nanda. Transaction Publishers.

Attfield, Robin. 2016. "Biocentrism." In *The International Encyclopedia of Ethics*, edited by Hugh LaFollette, 110. John Wiley & Sons. https://onlinelibrary.wiley.com/doi/epdf/10.1002/9781444367072.wbiee670.pub2?saml_referrer.

Kopnina, Helen and Haydn Washington. 2020. "Ethical Approaches to Conservation." In *Conservation: Integrating Social and Ecological Justice*, edited by Helen Kopnina and Haydn Washington, 17–31. Springer.

Taylor, Paul W. 1981. "The Ethics of Respect for Nature." *Environmental Ethics* 3 (3): 197–218.

Biodiversity

There are many definitions of biodiversity, but the most basic is simply the number of biological species present in a place. The term was first coined in 1980 (Lovejoy 1980; Norse and McManus 1980), but the idea dates to many earlier descriptions of ecosystems that emphasized the variety of plant and animal species present with considerable variation from place to place in those numbers, that is, in biodiversity.

Starting in the 1950s, ecologists argued that ecosystems with high biodiversity were more stable in the face of disturbances (e.g., weather events, human action) than systems with less biodiversity. Arguments about diversity and stability have generated many conceptual frameworks and theoretical models and a great deal of empirical research. An overarching finding might be "it depends" in the sense that there are many ways to conceptualize diversity and stability, and how they relate to each other may vary across types of ecosystems and kinds of stress placed on them (Hong et al. 2022). Nonetheless, there is a general sense that biodiversity is a desirable property for ecosystems if one values ecosystem stability. Part of the United Nations Sustainable Development Goal 15 is "halt biodiversity loss," and protecting biodiversity is seen as a key part of sustainability (→ Sustainable Development Goals) (Blicharska et al. 2019).

Aside from its relationship with ecosystem stability, biodiversity is seen by many as valuable in itself (→ Biocentrism) (→ Ecocentrism). For example, many see a tropical rainforest with myriad species as desirable and more valuable than an agricultural plantation dominated by one crop. One argument is based on the view that biodiversity is valuable to humans because biodiversity contributes to ecosystem services – the good things that ecosystems do for humans (→ Anthropocentrism) (→ Ecosystem Services). A diverse ecosystem provides many benefits that are lost when it is modified to produce a single crop even if the new system is stable (→ Feed Production). Another is that because the contributions of plants and animals to biodiversity have intrinsic value, they deserve ethical consideration beyond any contribution they make to human well-being (→ Biocentrism) (→ Ecocentrism). The value of biodiversity is being examined in detail in an ongoing international project, the Intergovernmental Science-Policy Platform on Biodiversity and Ecosystem Services, whose website is a good source of reports summarizing the latest knowledge on all aspects of these issues (IPBES 2019).

Biodiversity loss is driven by habitat transformation (such as conversion to agriculture), overexploitation (such as hunting and overfishing), pollution (such as toxics and increases in nitrogen), climate change, and biotic exchange (→ Habitat Loss) (→ Empty Forest) (→ Pollution) (→ Climate Change) (→ Invasive Animal Species) (Banks-Leite et al. 2020). The global threat of invasive species to biodiversity (Duenas et al. 2021) has led to calls for their elimination (→ Invasive Animal Species). For example, there are estimates that domestic cats in the US kill over 1.3 billion wild birds and over 6 billion wild mammals each year (Loss et al. 2013). But these estimates are controversial, and the debate over the issues has been contentious. Australia has identified feral and domestic cats as a major threat to endangered species and is engaged in a major program to reduce cat predation on wildlife (→ Wilderness Animals) (→ Mesopredator Release) (Dorph et al. 2024). But others question the estimated impacts of cats and raise ethical concerns about such programs including that they demean the intrinsic worth individual animals (Lynn et al. 2019).

References

Banks-Leite, Cristina, Robert Mark Ewers, Hollie Folkard-Tapp, and Adam Fraser. 2020. "Countering the Effects of Habitat Loss, Fragmentation, and Degradation through Habitat Restoration." *One Earth* 3: 672–676.

Blicharska, Malgorzata, Richard J. Smithers, Grzegorz Mikusiński, et al. 2019. "Biodiversity's Contributions to Sustainable Development." *Nature Sustainability* 2 (12): 1083–1093. https://doi.org/10.1038/s41893-019-0417-9.

Dorph, Annalie, G. Ballard, Sarah Legge, et al. 2024. "Current and Emerging Feral Cat Management Practices in Australia." *Wildlife Research* 51 (6): 1–13.

Duenas, Manuel-Angel, David J. Hemming, Amy Roberts, and Hilda Diaz-Soltero. 2021. "The Threat of Invasive Species to IUCN-Listed Critically Endangered Species: A Systematic Review." *Global Ecology and Conservation* 26: e01476.

Hong, Pubin, Bernhard Schmid, Frederik De Laender, et al. 2022. "Biodiversity Promotes Ecosystem Functioning Despite Environmental Change." *Ecology Letters* 25 (2): 555–569.

IPBES. 2019. *Global Assessment Report on Biodiversity and Ecosystem Services of the Intergovernmental Science-Policy Platform on Biodiversity and Ecosystem Services*, edited by E.S. Brondizio, J. Settele, S. Díaz, and H.T. Ngo. IPBES secretariat, Bonn, Germany. https://doi.org/10.5281/zenodo.3831673.

Loss, Scott R., Tom Will, and Peter P. Marra. 2013. "The Impact of Free-Ranging Domestic Cats on Wildlife of the United States." *Nature Communications* 4 (1): 1–8.

Lovejoy, Thomas E. 1980. "Forward." In *Conservation Biology: An Evolutionary-Ecological Perspective*, edited by Michael E. Soulé and B.A. Wilcox. Sinauer Associates.

Lynn, William S., Francisco Santiago-Ávila, Joann Lindenmayer, John Hadidian, Arian Wallach, and Barbara J. King. 2019. "A Moral Panic over Cats." *Conservation Biology* 33 (4): 769–776. https://doi.org/10.1111/cobi.13346.

Norse, Elliott A., and Roberta E. McManus. 1980. "Ecology and Living Resources: Biological Diversity." In *Environmental Quality 1980: The Eleventh Annual Report of the Council on Environmental Quality*, 31–80. Washington, DC: United States Council on Environmental Quality.

Biophilia

Biophilia combines two Greek terms, "bios" (life) and "philia" (affection). The term was created by Erich Fromm who defined biophilia as "the passionate love of life and of all that is alive" (Gunderson 2014a; Fromm 1973, 406). The term was popularized by E.O. Wilson who defines it as "the innate tendency to focus on life and lifelike processes" (Wilson 1984, 1). Fromm's argument derives from the Judeo-Christian tradition and emphasizes the love of others, human and non-human (→ Religion). Wilson's approach is grounded in natural history and the human fascination with other species. Both argue that, in contrast to human tendencies to exploit other species and place stress on the environment (→ Anthropocentrism), there are also strong tendencies to engage with, respect, and even love ecosystems and other species.

Several research themes have built on the idea of biophilia. Biophilic design in architecture attempts to create buildings that are sustainable in the sense of both minimizing impact on the environment and enhancing human well-being (Tekin et al. 2025). The approach takes account of the human love of nature and of non-human species in structuring the built environment. While Fromm and Wilson both assert that biophilia is a universal or at least a widespread human characteristic, that has not been tested empirically. Several lines of research touch on the psychology of biophilia. While not developed based on biophilia arguments, commonly used social psychological approaches to measuring values include biospheric altruism and concern for animals, both tapping aspects of biophilia (→ Altruism) (Dietz et al. 2017), and some scholars draw on the idea of biophilia to examine how individuals feel about specific environments, such as backyards (Woods et al. 2025). At a broader level, Gunderson (2014b) has argued for synthesizing biophilia as a motivation for sustainability efforts with consideration of social drivers such as the drive for profit.

References

Dietz, Thomas, Summer Allen, and Aaron McCright. 2017. "Integrating Concern for Animals into Personal Values." *Anthrozoös* 30 (1): 109–122.

Fromm, Erich. 1973. *The Anatomy of Human Destructiveness*. Fawcett Crest.

Gunderson, Ryan. 2014a. "Erich Fromm's Ecological Messianism: The First Biophilia Hypothesis as Humanistic Social Theory." *Humanity & Society* 38 (2): 182–204.

Gunderson, Ryan. 2014b. "Social Barriers to Biophilia: Merging Structural and Ideational Explanations for Environmental Degradation." *The Social Science Journal* 51 (4): 681–685. https://doi.org/10.1016/j.soscij.2014.06.002.

Tekin, Bekir Hüseyin, Gizem Izmir Tunahan, Zehra Nur Disci, and Hatice Sule Ozer. 2025. "Biophilic Design in the Built Environment: Trends, Gaps and Future Directions." *Buildings* 15 (14): 2516. https://doi.org/10.3390/buildings15142516.

Wilson, Edward O. 1984. *Biophilia*. Harvard University Press.

Woods, Vanessa, Lorrie Schmid, Hsuan Chen, and Melinda Knuth. 2025. "Backyard Biophilia: A Survey Instrument to Measure an Attraction to Biodiversity in the Home Garden." *HortScience* 60 (9): 1478–1485. https://doi.org/10.21273/HORTSCI18638-25.

Biopolitics

Biopolitics refers to the ways governments, institutions, and other authorities manage life, both human and non-human, through policies, practices, and systems that regulate bodies, populations, and environments. The term was coined by Michel Foucault to highlight how modern power operates not only through laws and punishment, but through the administration of health, reproduction, mobility, and survival itself (Foucault 1978/2025). Biopolitics draw attention to how life becomes an object of governance and how certain forms of life are valued, protected, or rendered expendable. Contemporary scholars have expanded biopolitical analysis to include animals, ecosystems, and multispecies futures, emphasizing that the power to regulate life extends beyond the human realm (→ Multispecies Futures) (Salzani and Cimatti 2024).

Biopolitics is particularly useful for understanding human relationships with non-human animals. Scholars have shown how states and institutions classify animals into categories such as livestock, wildlife, pets, liminal animals, or invasive species and govern them differently based on economic interests, risk management, or cultural norms (→ Livestock Production) (→ Wilderness Animals) (→ Liminal Animals) (→ Invasive Animal Species) (Braverman 2015). For example, farmed animals may be selectively bred, confined, or medicated to maximize efficiency, while wild animals may be monitored, culled, or relocated to maintain ecological or economic stability. Recent research demonstrates how industrial animal agriculture in countries such as India and China use biopolitical tools, including surveillance, selective breeding, and population control, to manage animal bodies at scale (→ Artificial Selection and Breeding) (Srinivasan 2017). Work in environmental humanities also shows how conservation programs decide which species receive care, which are left to decline, and how life endings are managed (van Dooren 2010). These examples reveal how animals' lives are shaped by political priorities rather than their own needs.

Biopolitics also provides an important lens for thinking about sustainability. Decisions about how to govern land, water, and multispecies populations determine who thrives and who is harmed in changing ecological conditions. Sustainable futures require rethinking biopolitical systems that render some species disposable while protecting others. Scholars argue that more inclusive, justice-oriented ecological governance can promote resilience by recognizing animal agency, ecological interdependence, and the uneven effects of environmental change (→ Agency) (→ Ecocentrism) (Lorimer 2015; Soares and Lorimer 2025). By examining how life is managed and valued, biopolitics encourages us to critically assess power structures shaping human-animal relations and to imagine sustainability models that support the flourishing of all species (→ Biocentrism).

References

Braverman, Irus. 2015. *Wild Life: The Institution of Nature*. Stanford University Press.

Foucault, Michel. 1978/2025. "The History of Sexuality Volume I." In *Feminist Studies: An Introductory Reader*, edited by Hemangini Gupta, Kelly Sharron, Carly Thomsen, and Abraham Weil. Routledge.

Lorimer, Jamie. 2015. *Wildlife in the Anthropocene: Conservation after Nature*. University of Minnesota Press.

Salzani, Carlo, and Felice Cimatti. 2024. *Biopolitical Animal*. Edinburgh University Press.

Soares, Filipa, and Jamie Lorimer. 2025. "Time for Rewilding: The Chronobiopolitics of Wildlife Conservation in Britain." *Environment and Planning E: Nature and Space* 8 (6): 1973–2001.

Srinivasan, Krithika. 2017. "Conservation Biopolitics and the Sustainability Episteme." *Environment and Planning A* 49 (7): 1458–1476.

van Dooren, Thom. 2010. "Pain of Extinction: The Death of a Vulture." *Cultural Studies Review* 16 (2): 271–289.

Biotechnology

Biotechnology is the human modification of biological processes to serve human ends. In our coevolution with dogs and later with other domesticated animals, humans applied selective breeding to improve characteristics desired by humans (→ Evolution) (→ Domestic Animals). Later, humans learned how to use fermentation and other biological processes to serve human ends. With the development of a deeper scientific understanding of molecular biology and especially of genetics, humans developed the ability to intervene much more deeply in the biology of plants, animals, and other organisms (→ Genetically Modified Organisms) (Tuhin 2025).

Most applications of biotechnology target plants, microbes, and other non-animal species. Some of these efforts are using biotechnology to improve the nutrient quality and reduce the environmental impact of plants grown for animal feed (→ Feed Production) (Vastolo et al. 2024). Some modifications of biotechnology to plants allow increased use of pesticides that can have adverse effects on animals (→ Genetically Modified Organisms). Biotechnology is also being applied to companion animals, both as part of veterinary medicine and in cloning individuals (Horowitz 2024). Cloning is also used in conservation efforts to help preserve endangered species, and there are ongoing projects to resurrect extinct species such as the woolly mammoth, the dodo, and the thylacine (Novak et al. 2025; Hearn 2025). The cloning of Dolly the sheep from adult somatic cells in 1996 was a major milestone in the potential application of biotechnology to conservation, and two recent successes in cloning endangered species have been achieved with the Przewalski's horse and the black-footed ferret (Novak et al. 2025).

Biotechnology is widely used to increase the quantity and quality of the production of food for human consumption and for making livestock animals more compatible with the conditions of raising animals in confined spaces (→ Livestock Production) (Petersen 2023). Twine (2010, 196) argues that the use of biotechnology in animal production "turns animal bodies into ethical biocapital" and perpetuates pre-existing global patterns of meat and dairy consumption, and Donovan (2022, 51) laments that any genetic engineering of animals as a violation of "their given natural identity and is therefore an affront to their inherent inalienable dignity."

References

Donovan, Josephine. 2022. *Animals, Mind, and Matter: The Inside Story*. Michigan State University Press.

Hearn, Timothy. 2025. "Can We Really Resurrect Extinct Animals, or Are We Just Creating Hi-Tech Lookalikes?" *Phys Org*, April 10. https://phys.org/news/2025-04-resurrect-extinct-animals-tech-lookalikes.html.

Horowitz, Alexandra. 2024. "Would You Clone Your Dog?" *The New Yorker*, June 24. www.newyorker.com/magazine/2024/07/01/would-you-clone-your-dog.

Novak, Ben J., Stewart Brand, Ryan Phelan, Sasha Plichta, Oliver A. Ryder, and Robert J. Wiese. 2025. "Towards Practical Conservation Cloning: Understanding the Dichotomy between the Histories of Commercial and Conservation Cloning." *Animals* 15 (7): 989. https://doi.org/10.3390/ani15070989.

Petersen, Bjoern. 2023. "Editorial: Genetic Engineering in Farm Animals." *Frontiers in Genetics* 14 (February): 1155201. https://doi.org/10.3389/fgene.2023.1155201.

Tuhin, Muhammad. 2025. "What Is Biotechnology? Everything You Need to Know About This Transformative Science." *Science News Today*, March 28. www.sciencenewstoday.org/what-is-biotechnology-everything-you-need-to-know-about-this-transformative-science.

Twine, Richard. 2010. *Animals as Biotechnology: Ethics, Sustainability and Critical Animal Studies*. Routledge.

Vastolo, Alessandro, Francesco Serrapica, Damiano Cavallini, Isa Fusaro, Alberto Stanislo Atzori, and Massimo Todaro. 2024. "Editorial: Alternative and Novel Livestock Feed: Reducing Environmental Impact." *Frontiers in Veterinary Science* 11 (July): 1441905. https://doi.org/10.3389/fvets.2024.1441905.

C

Capabilities Ethics

The capabilities approach to ethics argues that a just society will provide all sentient creatures (humans and other species) (→ Sentience) the capabilities that allow them to flourish. Nussbaum has argued "each sentient creature (capable of having a subjective point of view on the world and feeling pain and pleasure) should have the opportunity to flourish in the form of life characteristic for that creature" (2022, xxiv). Human capabilities include life, bodily health, bodily integrity, senses, imagination, thought, emotions, practical reason, affiliation with other humans, coexistence with other species, play, and control over one's environment. The capabilities approach suggests that all animals deserve ethical consideration, and coexistence with other species is a crucial human capability (→ Coexistence).

While a utilitarian approach to animal ethics (→ Utilitarianism) focuses on the aggregate pain and pleasure of sentient non-human species (and ethical choices are those that produce the greatest good for the greatest number), the capabilities approach gives weight to a broader spectrum of experience. Capabilities ethicists emphasize that ethical standing is not determined by similarity to human thinking and feeling but should be judged in terms of each species' own characteristic features.

Concerns with capabilities theory focus in part on the basis for deciding which species and which capabilities should be given consideration (Read and Birch 2023). The approach itself suggests that ethical analysis requires examination of what is required for flourishing, which differs across species and even within species depending on the circumstances of individuals. Ideas from Indigenous perspectives and from ecodemocracy and deliberative ethics might help articulate ways of assessing what is needed for the flourishing of non-human species (→ Ecodemocracy) (→ Indigenous and Decolonial Perspectives).

The capabilities approach grew directly from and has had influence on discussions of economic development (Chiappero-Martinetti 2020). Given the intent to consider the situation of non-human species and the broad consideration of what is needed for humans to flourish, the approach has also been engaged in discussions of sustainability. Some scholars suggest that sustainability and the need to consider non-human species would constrain

DOI: 10.4324/9781003528142-4

the unfettered use of all human capabilities (→ Sustainability) (Rauschmayer and Lessmann 2013). For example, consistent with capabilities ethics that gives each animal a flourishing life, Nussbaum has argued for an absolute ban on all killings of animals for sport, vanity, and unnecessary research (→ Animal Research) (Nussbaum 2012, 247).

References

Chiappero-Martinetti, Enrica, ed. 2020. *The Cambridge Handbook of the Capability Approach*. Cambridge University Press.

Nussbaum, Martha C. 2012. "The Capabilities Approach and Animal Entitlements." In *The Oxford Handbook of Animal Ethics*, 1st ed., edited by Tom L. Beauchamp and R.G. Frey, 228–252. Oxford University Press. https://doi.org/10.1093/oxfordhb/9780195371963.013.0009.

Nussbaum, Martha Craven. 2022. *Justice for Animals: Our Collective Responsibility*. Simon & Schuster.

Rauschmayer, Felix, and Ortrud Lessmann. 2013. "The Capability Approach and Sustainability." *Journal of Human Development and Capabilities* 14 (1): 1–5. https://doi.org/10.1080/19452829.2012.751744.

Read, Eva, and Jonathan Birch. 2023. "Animal Sentience and the Capabilities Approach to Justice." *Biology & Philosophy* 38 (4): 26. https://doi.org/10.1007/s10539-023-09914-0.

Captive Wild Animals

Captive wild animals are non-domesticated species, such as elephants, big cats, primates, marine mammals, and birds, who are held and managed in human controlled environments, including sanctuaries, research facilities, and zoos and aquariums (→ Sanctuaries) (→ Animal Research) (→ Zoos and Aquariums). Humans manage all aspects of captive wild animal lives including nutrient choice and intake, shelter, and breeding capacity. Thus, the ability of captive animals to make choices, their agency, is much constrained (→ Agency). Unlike domesticated animals, captive wild animals retain their species-typical behavioral, physiological, and ecological traits, even when raised in captivity (→ Domestic Animals). This creates complex welfare and ethical challenges as their biological needs often exceed what captive settings can provide, even under conditions designed to support enrichment, space, and social stimulation (→ Animal-Inclusive Design) (→ Welfare). Captive considerations require examining not only basic habitat requirements, but also the historical, cultural, and conservation rationales for holding wildlife captive for human use and human audiences (DiVincenti et al. 2023; Rose and Lewton 2025).

The complexity of keeping wild animals captive becomes clear when considering elephants. Elephants' well-documented cognitive sophistication and capacity for emotional experience mean that in captivity their welfare hinges on environments that provide ample space and the opportunity to maintain rich, stable social relationships consistent with their natural lives (→ Cognition). Individual elephant social histories, early-life experiences, and movement needs have long-term effects on the physical and psychological well-being of elephants in captivity even among the best facilities (Doyle et al. 2024). Similar dynamics appear with marine mammals, large cats, and some species of birds. Such complexities extend beyond typical welfare debates into the ethics of keeping wild animals.

Accredited zoos, aquariums, and other organizations keep animals as part of species survival plans, to educate the public, and for rehabilitation and habitat restoration efforts (→ Zoos and Aquariums) (Miranda et al. 2023), all of which are important for sustainability. However, the keeping of large wild charismatic megafauna raises concerns about energy use, land demands, genetic management, and the social and psychological impacts on these animals. Sustainable approaches to wildlife conservation increasingly call for compassionate conservation that balances captive breeding programs with ecosystem protection, community-based conservation, and policies that prioritize the long-term viability of species in their natural environments (→ Compassionate Conservation). In this sense, keeping wild animals in captivity illuminates both the possibilities and limits of human-driven conservation in an era of accelerating ecological crisis.

References

DiVincenti, Louis, Jr., Allen McDowell, and Elizabeth S. Herrelko. 2023. "Integrating Individual Animal and Population Welfare in Zoos and Aquariums." *Animals* 13 (10): 1577.

Doyle, Catherine, Heather Rally, Lester O'Brien, Mackenzie Tennison, Lori Marino, and Bob Jacobs. 2024. "Continuing Challenges of Elephant Captivity: The Captive Environment, Health Issues, and Welfare Implications." *PeerJ* 12: e18161.

Miranda, Rafael, Nora Escribano, María Casas, Andrea Pino-del-Carpio, and Ana Villarroya. 2023. "The Role of Zoos and Aquariums in a Changing World." *Annual Review of Animal Biosciences* 11 (1): 287–306.

Rose, Paul E., and Jack Lewton. 2025. "Key Concepts for Enhancing Zoo Animal Welfare: Coping, Comfort, Choice, Control, Challenge, and Compassion." *Journal of Applied Animal Welfare Science* 28 (3): 497–514.

Carnism

Carnism is the ideology that normalizes and justifies the eating of non-human animals, while also limiting reflection on why some species are valued as food and others are not. The term was popularized by social psychologist Melanie Joy (2020) in her framing of the background assumptions that must accompany the ideology as the 3Ns: normal, natural, and necessary. Some scholars have argued for a fourth N: nice (Piazza et al. 2015). The 3Ns/4Ns frame meat eating counter to vegetarianism and veganism as common sense rather than a contested ethical choice (→ Vegetarianism) (→ Veganism). Joy (2023) argues that carnism goes against the Golden Rule, to treat others as you would like to be treated, leading people to have disordered perceptions and a lack of empathy for non-human species (→ Empathy).

Carnism exists in the cultural narratives, institutional practices, religions, and subtle psychological processes designed to downplay non-human animal moral status and sentience and obscure the realities of livestock production conditions and environmental impacts (→ Religion) (→ Sentience) (→ Livestock Production). Research demonstrates that people rationalize eating meat by first prescribing fewer mental capacities to the animals they eat (Loughnan et al. 2010), and then by connecting the eating of animals to nature (it being natural), tradition (cultural or social), necessity (health related), or niceness (it tastes good or is "nice" to eat). The four justifications capture 83-91 percent of defenses for eating meat (Piazza et al. 2015). People who endorse the 4Ns are less likely to be motivated by ethical concerns in making food decisions, are less involved in animal welfare and advocacy, and more readily endorse speciesist attitudes (→ Welfare) (→ Advocacy) (→ Speciesism) (Piazza et al. 2015). Carnism is more than a set of eating habits; it functions as a durable social and cultural system that shapes how we see, feel about, and politically regulate certain animals to keep them classified as food.

Carnism has clear implications for sustainability. Livestock production contributes significantly to climate change, land degradation, water use, and pollution (→ Livestock Production) (→ Pollution). Global analyses show that animal-based foods carry higher environmental costs than plant-based foods (Poore and Nemecek 2018). Yet, sustainability is not just about emissions, but also about equity and well-being. Rural, low-income communities of color frequently bear disproportionate exposure to air and water pollution from livestock production, an issue often addressed in environmental justice literature (→ Environmental Justice). Understanding carnism through an ecological and social justice lens highlights that what is treated as "normal" or "necessary" in dominant food systems is not environmentally neutral, normal, or necessary.

References

Joy, Melanie. 2020. *Why We Love Dogs, Eat Pigs, and Wear Cows: An Introduction to Carnism*. Red Wheel.

Joy, Melanie. 2023. *How to End Injustice Everywhere: Understanding the Common Denominator Driving All Injustices, to Create a Better World for Humans, Animals, and the Planet*. Lantern Books.

Loughnan, Steve, Nick Haslam, and Brock Bastian. 2010. "The Role of Meat Consumption in the Denial of Moral Status and Mind to Meat Animals." *Appetite* 55 (1): 156–159.

Piazza, Jared, Matthew B. Ruby, Steve Loughnan, et al. 2015. "Rationalizing Meat Consumption. The 4Ns." *Appetite* 91: 114–128.

Poore, Joseph, and Thomas Nemecek. 2018. "Reducing Food's Environmental Impacts through Producers and Consumers." *Science* 360 (6392): 987–992. https://doi.org/10.1126/science.aaq0216.

Citizen Science

Citizen science is the co-production of scientific knowledge by researchers and non-professional volunteers who help with data collection and other tasks, such as species identification, data annotation, and classification (Criscuolo et al. 2023). Scientific collaboration between researchers, research institutions, and the public has been linked to the democratization of science and encourages communication and cooperation among different disciplines and research communities (Criscuolo et al. 2023). Citizen science is also used to describe the collection of the public's own evidence to shape policies and increase community understanding, including participatory projects where citizen scientists help define the topics, collaborate on the design, or even manage the entire research project themselves (Criscuolo et al. 2023).

Citizen science is particularly useful in monitoring changes in the environment and wildlife diversity over large geographic areas. One of the longest running citizen science projects is the Christmas bird count in the US that has been collecting data on birds since 1900. Similar citizen science networks in Ireland have been gathering data on the number and diversity of birds since the 1960s, and recent initiatives that support Ireland's commitment to nature conservation include Nature Watch and Greenwave (the recording of natural events every Spring by primary school students such as bird migration and frogspawn) (Donnelly et al. 2014). On a global scale, eBird is a worldwide online checklist for bird watchers to record the birds they see, share data, and collect information on bird abundance and distribution at large dimensions of time and place (Donnelly et al. 2014). The phone app iNaturalist provides species identification of photos submitted and aggregates the identification and spatial location into a database that is proving useful for research (Mason et al. 2025).

In the marine environment, some of the most acclaimed citizen science projects are devoted to collecting data on marine animal forests (→ Animal Forests) (Zorrilla-Pujana 2020). For example, Reef Check and Coral Watch collect data on coral reefs around the world, REEF (Reef Environmental Education Foundation) has hands-on programs such as the Invertebrate and Algae Monitoring Program that engages local communities in conservation activities and the Marine Forests Initiative that is focused on data collection on benthic organisms, macroalgae, and plants by citizen scientists who record their observations of forest-forming species on worldwide maps (Zorrilla-Pujana 2020).

Citizen science projects have also been used to engage ecotourists in monitoring local marine biodiversity (Dionisio et al. 2022), monitor and mitigate roadkill in Australia (Englefield et al. 2020), assess climate change shifts in desert reptile communities (Barrows et al. 2016), and collect vector-borne arthropods (ticks, fleas) from animal hosts to monitor zoonotic diseases (→ Ecotourism) (→ Road Ecology) (→ Climate Change) (→ Zoonosis) (Poh et al. 2022). Citizen science studies of the diversity of wild and managed bees have contributed to 15 of the 17 Sustainable Development Goals (such as Life on land, Sustainable Cities and Communities, and Responsible Consumption) and at least 30 relevant Sustainable Development Targets (→ Sustainable Development Goals) (Koffler et al. 2021).

Finally, citizen science approaches to the production of knowledge can address environmental justice concerns by engaging marginalized communities in scientific research and "amplify the voices of communities disproportionately affected by environmental degradation and pollution … [thus] leading to equitable and sustainable outcomes" (→ Environmental Justice) (→ Intersectional Environmentalism) (Bhandari 2024, 2). Often marginalized communities have perspectives that are essential for applying science to local problems (Dietz 2023). By engaging the public, citizen science generates valuable data, enhances public engagement with science, and empowers communities to advocate for environmental sustainability and a healthy planet (Bhandari 2024).

References

Barrows, Cameron W., Josh Hoines, Michael S. Vamstad, Michelle Murphy-Mariscal, Kristen Lalumiere, and James Heintz. 2016. "Using Citizen Scientists to Assess Climate Change Shifts in Desert Reptile Communities." *Biological Conservation* 195: 82–88.

Bhandari, Medani P. 2024. "Citizen Science and Its Applicability for Sustainability and a Healthy Planet." *Academia Environmental Sciences & Sustainability* 1: 1–14. https://doi.org/10.20935/AcadEnvSci7270.

Criscuolo, Laura, Alba L'Astorina, René van der Wal, and Laura Colucci Gray. 2023. "Recent Contributions of Citizen Science on Sustainability Policies: A Critical Review." *Current Opinion in Environmental Science & Health* 31: 100423.

Dietz, Thomas. 2023. *Decisions for Sustainability: Facts and Values*. Cambridge University Press.

Dionisio, Mara, Mafalda Mendes, Marc Fernandez, Valentina Nisi, and Nuno Nunes. 2022. "Aqua: Leveraging Citizen Science to Enhance Whale-Watching Activities and Promote Marine-Biodiversity Awareness." *Sustainability* 13: 14203. https://doi.org/10.3390/su142114203.

Donnelly, Alison, Olivia Crowe, Eugenie Regan, Sinead Begley, and Amelia Caffarra. 2014. "The Role of Citizen Science in Monitoring Biodiversity in Ireland." *International Journal of Biometeorology* 58: 1237–1249.

Englefield, Bruce, Melissa Starling, Bethany Wilson, Caidyrn Roder, and Paul McGreevy. 2020. "The Australian Roadkill Reporting Project—Applying Integrated Professional Research and Citizen Science to Monitor and Mitigate Roadkill in Australia." *Animals* 10: 1112. https://doi.org/10.3390/ani10071112.

Koffler, Sheina, Celso Barbiéri, Natalia P. Ghilardi-Lopes, et al. 2021. "A Buzz for Sustainability and Conservation: The Growing Potential of Citizen Science Studies on Bees." *Sustainability* 13: 959. https://doi.org/10.3390/su13020959.

Mason, Brittany M., Thomas Mesaglio, Jackson Barratt Heitmann, et al. 2025. "iNaturalist Accelerates Biodiversity Research." *BioScience* 75 (11): 953–965. https://doi.org/10.1093/biosci/biaf104.

Poh, Karen C., Jesse R. Evans, Michael J. Skvarla, and Erika T. Machtinger. 2022. "All for One Health and One Health for All: Considerations for Successful Citizen Science Projects Conducting Vector Surveillance from Animal Hosts." *Insects* 13: 492.

Zorrilla-Pujana, J. 2020. "Advances in the Marine Animal Forests Scientific Outreach and Citizen Science." In *Perspectives on the Marine Animal Forests of the World*, edited by Sergio Rossi and Lorenzo Bramanti, 515–530. Springer Nature, Switzerland.

Climate Change

Human caused (anthropogenic) climate change is one of the most serious of the global environmental changes that are impacting humans, other species, and the prospects for sustainability (→ Global Environmental Change). The climate of the Earth has shifted over time as a result of the release of greenhouse gases through volcanic activity, shifts in continents that change circulation of heat and moisture via oceans and wind, and changes in the output of radiation from the sun. Some of these climate shifts have triggered major extinctions (→ Extinction).

Human actions have an impact on the climate. European colonization of the Americas led to massive shifts in land use as Indigenous people were impacted by huge increases in mortality and by displacement from their traditional lands leading to the abandonment of agriculture (→ Colonialism). This process may have led to new vegetation absorbing CO_2 (carbon dioxide) on a large enough scale that it contributed to a "little ice age" from the fourteenth to the nineteenth centuries (Koch et al. 2019). Since the late eighteenth century, the increased burning of fossil fuels (coal, oil, natural gas) has led to a 50 percent increase in the concentration of greenhouse (heat trapping) gases in the atmosphere and a warming such that the years 2023, 2024, and 2025 were the warmest years in the historical record (National Academies of Sciences, Engineering, and Medicine 2025).

The effects of recent and anticipated climate changes on non-human animals are massive and complex. Climate change is a major influence on the loss of biodiversity, with impacts that unfold in many ways (→ Biodiversity) (Pörtner et al. 2021). Warming and associated changes in weather and seasonal patterns influence species ranges. This often results in expanded ranges for some species such as coyotes making them invasive and leading to the loss of native species (→ Native Species) (→ Invasive Animal Species). Disease carrying species are able to move into new areas as winters become milder, increasing risk for domestic and wild animals (→ Zoonosis) (Greening et al. 2025). Climate change combined with land use change will cause substantial loss of suitable habitat for many animals that could lead to the extinction of 8000 vertebrate species (→ Habitat Loss) (Vardi et al. 2025). Changing climate can also disrupt patterns of interactions between plants and animals, for example by shifting the timing of hatchings of pollinators and blooming of plants that impact ecosystems and global sustainability (Martén-Rodríguez et al. 2025).

References

Greening, Sabrina S., Lucie R. Pascarosa, Avery L. Munster, Roderick B. Gagne, and Julie C. Ellis. 2025. "Climate Change as a Wildlife Health Threat: A Scoping Review." *BMC Veterinary Research* 21 (1): 60. https://doi.org/10.1186/s12917-025-04516-2.

Koch, Alexander, Chris Brierley, Mark M. Maslin, and Simon L. Lewis. 2019. "Earth System Impacts of the European Arrival and Great Dying in the Americas after 1492." *Quaternary Science Reviews* 207: 13–36.

Martén-Rodríguez, Silvana, Edson Jacob Cristobal-Pérez, Martín Hesajim de Santiago-Hernández, et al. 2025. "Untangling the Complexity of Climate Change Effects on Plant Reproductive Traits and Pollinators: A Systematic Global Synthesis." *Global Change Biology* 31 (2): e70081. https://doi.org/10.1111/gcb.70081.

National Academies of Sciences, Engineering, and Medicine. 2025. *Effects of Human-Caused Greenhouse Gas Emissions on U.S. Climate, Health, and Welfare*. Washington, DC: The National Academies Press. https://doi.org/10.17226/29239.

Pörtner, Hans-Otto, Robert J. Scholes, John Agard, et al. 2021. "Scientific Outcome of the IPBES-IPCC Co-Sponsored Workshop on Biodiversity and Climate Change", Version 5. Bonn, Germany: IPBES Secretariat. https://doi.org/10.5281/ZENODO.4659158.

Vardi, Reut, Gopal Murali, Gabriel Henrique de Oliveira Caetano, Uri Roll, and Shai Meiri. 2025. "Effects of Future Climate Extreme Heat Events and Land Use Changes on Land Vertebrates." *Global Change Biology* 31 (12): e70625. https://doi.org/10.1111/gcb.70625.

Coexistence

Coexistence is the dynamic, enduring state in which animals and humans coadapt to living in a shared landscape (König et al. 2020). Coexistence requires the tolerance and effective management of conflict and risks for humans (attacks on people and livestock, destruction of crops) and for wildlife (illegal killing, destruction of habitat) (→ Conflict with Terrestrial Animals) (→ Conflict with Ocean Animals) (→ Wildlife Crime) (Pooley et al. 2020). Apex carnivores such as wolves and dingoes are in conflict with humans but are so essential to the ecosystem that multiple non-lethal strategies have been devised to coexist with them. For example, predator-smart farming controls both wolf and dingo predation on livestock through guardian dogs and donkeys, fences, and the removal of attractants such as rotting carcasses (→ Livestock Production).

One of the best-known examples of human-wildlife coexistence takes place in the Gir Forest of India. The lions of the Gir Forest have learned behaviors that have sustained their species in a heavily dominated human landscape (→ Learning for Survival). In the nineteenth century the Gir lions hunted and ate goats, cows, buffaloes, and humans in and beyond the forest, but in 1965 lions were declared a protected species and lion-human relations took a turn for the better. The Gir lions now live on a mix of 25 percent domestic livestock and 75 percent wild prey and rarely attack humans who live in communities surrounding the forest (Yadav and Agarwal 2021). Both lions and human residents of the area developed a pattern of behaviors ranging from avoidance to tolerance to coexistence, and the lions behave differently from lions anywhere else: they have adapted to humans, coming "as close to trusting people as wild predators can" (Rangarajan 2013, 126).

California sea lions have also learned to adapt to a human-dominated world. Once in decline and on the brink of extinction, the California sea lion is one of the most successful recoveries in the modern conservation era (Schakner and Blumstein 2021). In addition to the protections provided by the Marine Mammal Protection Act (1972), the recovery of the California sea lion is based on their capacity to behaviorally innovate, learn, and transmit knowledge of novel food sources through social networks (→ Marine Mammal Protection Act) (→ Learning for Survival) (→ Cultural Transmission) (Schakner and Blumstein 2021). River otters in North America have also recovered after human-induced urbanization and pollution degraded their habitats, and they now coexist with humans in anthropogenic landscapes, particularly docks. River otters on Protection Island, British Columbia, are attracted to docks because they are close to water and provide opportunities to quickly evade threats and to access foraging opportunities; human activities do not impede the otters' use of the space for resting and social activities (Lesage et al. 2025).

Pooley et al. (2020) argue that coexistence research should not be dominated by conflict studies of human-wildlife interactions, but rather guided by historical research, ethnographies, and research on human-animal relationships (→ Conflict with Terrestrial Animals) (→ Conflict with Ocean Animals). Sustainable human-wildlife coexistence depends on transdisciplinary science-stakeholder approaches in the implementation of policy and management strategies (König et al. 2020). Finally, the wellbeing of both humans and wildlife is inextricably linked in reciprocal relationships, and the integration of ecological, physical, and social sciences is needed to understand and promote coexistence and long-term sustainability (→ Coupled Human and Natural Systems) (Carter et al. 2019).

References

Carter, Neil H., Jeremy T. Bruskotter, John Vucetich, et al. 2019. "Towards Human–Wildlife Coexistence through the Integration of Human and Natural

Systems: The Case of Grey Wolves in the Rocky Mountains, USA." In *Human–Wildlife Interactions: Turning Conflict into Coexistence*, edited by Beatrice Frank, Jenny A. Glikman, and Silvio Marchini, 384–413. Cambridge University Press.

König, Hannes J., Christian Kiffner, Stephanie Kramer-Schadt, Christine Fürst, Oliver Keuling, and Adam T. Ford. 2020. "Human-Wildlife Coexistence in a Changing World." *Conservation Biology* 34 (4): 786–794.

Lesage, Caroline E., Angela Kross, and Sarah E. Turner. 2025. "North American River Otter Habitat Suitability and Human-Wildlife Coexistence on Protection Island, British Columbia, Canada." *Journal of Wildlife Management* 89: e22719.

Pooley, Simon, Saloni Bhatia, and Anirudhkumar Vasava. 2020. "Rethinking the Study of Human-Wildlife Coexistence." *Conservation Biology* 35 (3): 784–793.

Rangarajan, Mahesh. 2013. "Animals with Rich Histories: The Case of the Lions of Gir Forest, Gujarat, India." *History and Theory* 52: 109–127.

Schakner, Zachary A., and Daniel T. Blumstein. 2021. "The California Sea Lion: Thriving in a Human-Dominated World." In *Ethology and Behavioral Ecology of Otariids and the Odobenid. Ethology and Behavioral Ecology of Marine Mammals*, edited by C. Campagna, and R. Harcourt, 347–365. Springer. https://doi.org/10.1007/978-3-030-59184-7_18.

Yadav, Pushpa, and Punit Kumar Agarwal. 2021. "Co-Existence of Asiatic Lion in Human Vicinity: Sustaining the Roar of Kingdome Outside the Gir Forest." *Environment and Ecology* 39 (4A): 1178–1181.

Cognition

Cognition refers to the mental processes that allow animals (human and non-human) to perceive, learn, solve problems, remember, communicate, and make decisions. In humans, cognition is often referred to as human intelligence. For non-human animals, cognition is thought of as a set of adaptive strategies that help individuals navigate challenges such as finding food, avoiding predators, coordinating social behavior, and raising offspring (→ Cultural Transmission) (→ Learning for Survival) (Shettleworth 2009).

Early philosophical debates about animal cognition and sentience were shaped heavily by René Descartes, who argued in the seventeenth century that non-human animals were essentially "automata" or mechanical beings incapable of genuine feeling or consciousness (→ Sentience). This view stood in contrast to other thinkers, such as Montaigne, who pointed to animals' emotional responsiveness and social behavior as evidence of inner experience and is also in sharp contrast with many Indigenous views that ascribe intelligence and agency to non-humans and consider them part of the community (→ Indigenous and Decolonial Perspectives). Contemporary research demonstrates that many species engage in sophisticated forms of information processing shaped by their ecological and social environments. The study of animal cognition resides at the intersection of psychology, ethology, neurobiology, and evolutionary theory and is generally focused on understanding how animals think and how mental capacities emerge (→ Cultural Transmission) (→ Evolution).

Ongoing research suggests that most non-human animals have a range of cognitive abilities, many of which we know very little about. One influential line of cognition research focuses on problem solving skills. Birds such as crows, ravens, and jays can use and even modify tools to access hard-to-reach food, demonstrating cognitive flexibility once thought limited to primates (Jacobs and Osvath 2023). Similarly, elephants have been shown to coordinate actions and understand when cooperation is required to complete a complex task (Plotnik and de Waal 2014; Goldenberg and Wittemyer 2020). This is an indication not only of social cognition, but also of intentionality and agency (→ Agency). Collectively, this body of literature shows how cognition is shaped by a species' evolutionary history and highlights the importance of examining species-specific cognition.

Understanding non-human animal cognition has increasing relevance for sustainability. Recognizing animals as thinking, decision-making beings can influence how societies design conservation programs, manage habitats, and define ethical responsibilities toward other species. For example, acknowledging that many animals rely on memory, learning, culture and/or long-term social bonds helps explain why habitat fragmentation or climate-driven disruptions can have cognitive and emotional consequences (→ Habitat Loss). Research on animal cognition also informs coexistence strategies by identifying how animals interpret human actions, navigate landscapes, or adapt to environmental change (→ Coexistence). By incorporating cognitive science into sustainability efforts, conservationists can create policies that respect animals' capacities and promote ecological resilience grounded in an appreciation of non-human animal minds.

References

Goldenberg, Shifra Z., and George Wittemyer. 2020. "Elephant Behavior Toward the Dead: A Review and Insights from Field Observations." *Primates* 61 (1): 119–128.

Jacobs, Ivo, and Mathias Osvath. 2023. "Tool Use and Tooling in Ravens (*Corvus corax*): A Review and Novel Observations." *Ethology* 129 (3): 169–181.

Plotnik, Joshua M., and Frans B.M. de Waal. 2014. "Asian Elephants (Elephas maximus) Reassure Others in Distress." *PeerJ* 2: e278.

Shettleworth, Sara J. 2009. *Cognition, Evolution, and Behavior.* Oxford University Press.

Colonialism

Colonialism refers to the political, economic, and cultural domination exerted by one society over another through practices of territorial takeover, resource extraction, and institutional restructuring. Although often discussed as a historical process, scholars increasingly emphasize colonialism as an ongoing system that shapes contemporary inequalities, environmental relationships, and global patterns of consumption (Whyte 2017). Colonial rule reorganizes land tenure, disrupts Indigenous governance, and positions extractive economies as central to development, leading to long-term social and ecological consequences. It can also introduce new zoonotic diseases to Indigenous populations, often with devastating results that multiply the other adverse effects. Understanding colonialism requires attention not only to past conquests, but also to the enduring structures and worldviews that continue to inform current environmental and political systems (→ Indigenous and Decolonial Perspectives).

These dynamics also extend to non-human animals. Colonial administrations routinely introduce livestock species, reshape hunting regulations, and implement conservation practices that undermine Indigenous relationships with local ecologies. For example, in the late nineteenth century, US policymakers and military officials endorsed the large-scale killing of buffalo as a deliberate strategy to weaken Plains Indigenous nations and pressure them onto reservations. Because the buffalo was central to Indigenous economies' mobility and political autonomy, their destruction targeted the foundation of everyday life and made communities more vulnerable to federal control (Isenberg 2020). Scholars have documented that military leaders such as General Philip Sherida supported commercial hide hunters because eliminating the herds would deprive Indigenous groups of food and material resources thereby limiting their ability to resist US expansion (Smits 1994). The resulting near-extinction of the species functioned as both an ecological catastrophe and a calculated form of state violence aimed at undermining Indigenous sovereignty (→ Extinction).

Recent research shows that colonial interventions often transformed animals into instruments of economic expansion through ranching, fur markets, and commercial hunting, while simultaneously dispossessing Indigenous communities of their stewardship practices. A contemporary example can be seen in debates over invasive species management, where some species introduced through colonial trade routes, such as pigs or goats on Pacific islands, are now subject to eradication campaigns that could disproportionately impact Indigenous livelihoods and cultural relationships with the land (→ Invasive Animal Species) (Wehi et al. 2023; Reo et al. 2017). These cases illustrate how non-human animals become entangled in colonial projects and how these legacies persist in present-day society and wildlife policy.

Scholars argue that many global sustainability challenges, including biodiversity loss, climate vulnerability, and inequitable conservation practices, cannot be understood without addressing the colonial histories that shaped land-use patterns and resource governance (→ Biodiversity) (→ Climate Change) (Buscher and Fletcher 2020). Conservation initiatives built on colonial models often prioritize protecting wildlife for tourism or global markets rather than supporting the reciprocal, place-based ecological practices long maintained by Indigenous peoples (→ Wildlife Tourism) (→ Wildlife Trade). Moving toward sustainable futures requires confronting these historical structures, centering Indigenous knowledge systems, and acknowledging how colonial land regimes continue to shape human-animal relationships and sustainable policy.

References

Buscher, Bram, and Robert Fletcher. 2020. *The Conservation Revolution: Radical Ideas for Saving Nature Beyond the Anthropocene*. Verso Books.

Isenberg, Andrew C. 2020. *The Destruction of the Bison: An Environmental History, 1750–1920*. Cambridge University Press.

Reo, Nicholas J., Kyle Whyte, Darren Ranco, Jodi Brandt, Emily Blackmer, and Braden Elliott. 2017. "Invasive Species, Indigenous Stewards, and Vulnerability Discourse." *American Indian Quarterly* 41 (3): 201–223.

Smits, David D. 1994. "The Frontier Army and the Destruction of the Buffalo: 1865–1883." *Western Historical Quarterly* 25 (3): 312–338.

Wehi, Priscilla M., Katie L. Kamelamela, Kyle Whyte, Krushil Watene, and Nicholas Reo. 2023. "Contribution of Indigenous Peoples' Understandings and Relational Frameworks to Invasive Alien Species Management." *People and Nature* 5 (5): 1403–1414.

Whyte, Kyle. 2017. "Indigenous Climate Change Studies: Indigenizing Futures, Decolonizing the Anthropocene." *English Language Notes* 55 (1): 153–162.

Commons Dilemma

The term "commons" refers to a resource, such as grazing land or a forest, that is not owned by any individual or organization, but where all members of a community have open access to use it. A commons dilemma occurs when such an open access resource might be overused and degraded because of, for example, grazing too many sheep, cutting too much firewood, or overfishing (Ostrom et al. 2002; McCay and Acheson 1990). If each member of the community limits their use of the resource, then the resource can be maintained without degrading it. But everyone faces the dilemma. If they limit their use of the resource but others do not, the common resource will still collapse, and they as an individual have limited the benefits they receive from using the resource when it is still viable. So, a narrowly self-interested person will not limit their use, and if everyone does this, the common resource will be degraded. Of course, this analysis is anthropocentric in that only the use of the resource for human ends is considered (→ Anthropocentrism). But when the resource is an ecosystem, both biotic and abiotic elements will be impacted by its condition and suffer from degradation (→ Ecocentrism). Many of the most prominent examples of commons, such as fisheries or grazing lands, involve the human use of other animal species and ecosystems (→ Livestock Production) (→ Aquaculture).

This is sometimes called "the tragedy of the commons" because it is assumed that under common ownership and control, collapse of the resource is inevitable. But people often act altruistically rather than with narrow self-interest, including taking account of the well-being of other humans and other species (→ Altruism). The historical evidence suggests that while some commons do collapse from overexploitation, there are also many cases where they are managed by communities for decades or centuries without being overexploited (→ Indigenous and Decolonial Perspectives). So many scholars prefer to call the situation the "drama of the commons," taking account of both tragic and positive outcomes to commons management.

While the theory of the commons was developed primarily around local resources, the same logic applies to systemic global environmental change, such as climate change (→ Climate Change) (Dietz et al. 2003). The actions of one individual or even one nation will not change the trajectory of climate change, and if everyone acts according to narrow self-interest, then climate change is inevitable. But in fact, individuals and nations often act for the common good, including coordinating with others to reach agreements about how to successfully manage common global resources (Ostrom 2010).

The challenge of managing commons arises in nearly every aspect of sustainability. Actions by a community to provide services, public education or police or fire protection, face the same problem of needing to get most people to cooperate in supporting them. Finally, some research shows that other animal species can successfully manage a commons but also face difficulties in doing so, such as chimpanzees who have been documented to overcome the tragedy of the commons in collectively sustaining a renewable, collapsible common resource (juice and carrots) in an experimental setting (Koomen and Herrmann 2018).

References

Dietz, Thomas, Elinor Ostrom, and Paul C. Stern. 2003. "The Struggle to Govern the Commons." *Science* 301 (5652): 1907–1912.

Koomen, Rebecca, and Esther Herrmann. 2018. "Chimpanzees Overcome the Tragedy of the Commons with Dominance." *Scientific Reports* 8 (1): 10389. https://doi.org/10.1038/s41598-018-28416-8.

McCay, Bonnie J., and James M. Acheson. 1990. *The Question of the Commons: The Culture and Ecology of Communal Resources*. University of Arizona Press.

Ostrom, Elinor. 2010. "Polycentric Systems for Coping with Collective Action and Global Environmental Change." *Global Environmental Change* 20: 550–557.

Ostrom, Elinor, Thomas Dietz, Nives Dolsak, Paul C. Stern, Susan Stonich, and Elke Weber. 2002. *The Drama of the Commons*. National Academy Press.

Compassionate Conservation

Compassionate conservation is an emerging approach in conservation science that emphasizes the moral significance of individual animals alongside broader ecological goals. Developed in response to practices that prioritize species or ecosystem outcomes at the expense of individual well-being, compassionate conservation argues that conservation actions should be guided by principals such as "first, do no harm," the recognition of individual animals as deserving moral consideration, and the integration of ethics into ecological management. Rather than treating animals as interchangeable units in a population, this framework seeks to reconcile biodiversity protection with respect for the lives, experiences, and intrinsic value of individual animals (Coghlan and Cardilini 2022). The term was coined by Marc Bekoff (2010), an American biologist, ethologist, and behavioral ecologist.

In practice, compassionate conservation reshapes how managers respond to human-wildlife conflicts, invasive species, and wildlife population control (→ Conflict with Terrestrial Animals) (→ Conflict with Ocean Animals), (→ Invasive Animal Species) (→ Wilderness Animals) (Lynn et al. 2025). For example, debates over the lethal culling of introduced species such as feral horses in Australia highlight tensions between traditional conservation strategies and compassionate conservation principles (→ Feral Animals). Critics of culling argue that lethal methods disregard the sentience and social complexity of targeted animals and may cause unnecessary suffering (→ Sentience) (→ Welfare) (Wallach et al. 2018). Instead, compassionate conservationists advocate for non-lethal approaches, including fertility control, translocation, coexistence, and habitat modification. These alternatives aim to address ecological concerns while minimizing harm to individual animals, reflecting the perspective that conservation should not treat living beings as disposable tools to achieve environmental goals.

Sustainable practices require not only maintaining functional ecosystems but also fostering coexistence between humans and non-human species in ways that reduce conflict and promote ecological integrity (→ Coexistence). By centering compassion and harm reduction, compassionate conservation encourages approaches that align ecological restoration, animal welfare, and social values, supporting more durable and publicly acceptable conservation outcomes (→ Restoration Ecology) (→ Welfare) (→ Values) (Vucetich et al. 2015). Compassionate conservation offers a framework for integrating ethical consideration of animals with broader goals of sustaining biodiverse, healthy, and just socio-ecological systems.

References

Bekoff, Marc. 2010. "Conservation Lacks Compassion." *New Scientist* 207 (2775): 24–25.

Coghlan, Simon, and Adam P.A. Cardilini. 2022. "A Critical Review of the Compassionate Conservation Debate." *Conservation Biology* 36 (1): e13760.

Lynn, William S., Liv Baker, William T. Borrie, et al. 2025. "Compassionate Conservation in Practice: A Values-Driven, Interdisciplinary, Pluralistic, and Deliberative Community." *Biological Conservation* 303 (11100): 2.

Vucetich, John A., Jeremy T. Bruskotter, and Michael Paul Nelson. 2015. "Evaluating Whether Nature's Intrinsic Value Is an Axiom of or Anathema to Conservation." *Conservation Biology* 29 (2): 321–332.

Wallach, Arian D., Marc Bekoff, Chelsea Batavia, Michael Paul Nelson, and Daniel Ramp. 2018. "Summoning Compassion to Address the Challenges of Conservation." *Conservation Biology* 32 (6): 1255–1265.

Conflict with Ocean Animals

Conflict between humans and ocean animals occurs when the actions of humans (such as fishing, aquaculture, recreation) or by ocean animals (such as stingray stings, competition over shared resources) have an adverse effect on the other, or there exist threats (actual or perceived) by ocean animals to human property, security, recreation, or safety (Guerra 2019). Human impacts on the ocean over time have resulted in substantial declines in marine predators and megafauna and cascading ecological change (→ Trophic Cascades). Conservation efforts, such as the moratorium on whaling (1984), the protection of North Atlantic fur seals (1911), the Endangered Species Act (1973), and the Marine Mammal Protection Act (1972), have increased many ocean populations of cetaceans (e.g., the humpback whale) and pinnipeds (e.g., the California sea lion) (→ Endangered Species Act) (→ Marine Mammal Protection Act) (NOAA Fisheries 2022).

However, the global abundance of elasmobranchs (sharks and rays) has declined by 71 percent because of overfishing, making them the most threatened vertebrate lineage after amphibians (Pacoureau et al. 2021). The collapse of the oceanic shark population will have negative consequences for economic and ecological systems, and immediate steps are imperative for long-term sustainability of "some of the most iconic and functionally important animals in our oceans" (Pacoureau et al. 2021, 57). However, the shark is one of the most feared marine predators, a fear perpetuated by inflammatory media that began 50 years ago with the film *Jaws* and continues today with films such as *The Shallows* and *Shark Night*. However, sharks pose little risk to humans. For example, surfers in California have a one in 17,000,000 chance of a shark bite, but the fear continues unabated – public perception of danger and perceived conflict are just as important as actual events (Guerra 2019).

Although uncommon at only 75-100 incidents per year globally, shark bites are one of the most widely reported types of human-wildlife conflict (Huveneers et al. 2024). Mitigation strategies are needed to increase protection of people while minimizing harm to sharks and other marine animals (Martin et al. 2022). In a community study of the acceptance of shark mitigation strategies in Australia, non-lethal approaches, such as drone surveillance, helicopter detection, public notifications, and education about sharks, were preferred over invasive or lethal approaches such as shark nets (→ Coexistence) (Martin et al. 2022). Shark nets are large mesh nets that are designed to entangle sharks who swim into them by chance, but they also entangle other marine animals including whales, dolphins, turtles, and seals. Of particular importance in the mitigation of shark bites is the personal responsibility that humans must assume when entering the ocean such as avoiding the water at times of high risk and using a personal deterrent (Huveneers et al. 2024).

Sharks (along with toothed whales) are also the primary marine animals involved in one of the oldest human-animal conflicts, depredation, or the direct feeding by large marine predators on fish caught on fishing gear. For human fishers, the cost of depredation is primarily economic including reduced catch and damage to equipment; for the depredating species, the costs are increased risk of injury and lethal retaliation, and the alternation of trophic interactions in ecosystems by facilitated access to prey that change predator foraging behavior and diet (→ Trophic Cascades) (Mollier et al. 2024). The conflict between marine predators and commercial fisheries is particularly problematic. In Australian longline fishery, 20 percent of catch is lost to shark depredation, and killer whale depredation in Alaskan longline fisheries has reduced overall catch by as much as 28 percent (Guerra 2019). Attempts to reduce depredation risk include modifications in angler behavior and gear (such as landing fish quickly and using heavier tackle) (Drymon et al. 2024), the use of non-lethal deterrence devices (acoustic and explosives/gunshots), avoidance of high depredation times and/or areas, electronic monitoring (→ Coexistence), and (although rare) legal selective killing of perpetrators (Tixler et al. 2021). Mitigation strategies must be interdisciplinary and address multiple stakeholders, including the individual animal species and human stakeholders (conservationists, fishers, and policymakers) during both planning and implementation processes (Guerra 2019).

Another human-marine animal conflict is the negative impacts of whale watching tourism on cetaceans (→ Wildlife Tourism). For example, humpback whales are threatened by whale watching vessel traffic during their breeding season in the designated Marine Protected Area in Bahía de Banderas, Mexico. Cetacean density in Bahía de Banderas is 15 times greater than in the surrounding open waters because of the habitat's ideal conditions for mating, giving birth, and raising offspring (Noriega et al. 2023). Humpback whales return to the area every year at the same time, creating a popular ecotourism activity of whale watching, but the increased boat traffic from whale watching adversely impacts whale health and behavior, including changes in swim speed, directions, and collisions with boats (Noriega et al. 2023). Some mitigation strategies include traffic rerouting, speed reduction, and the identification and monitoring of areas with high vessel density that threaten the whale population (→ Coexistence) (Noriega et al 2023). Whales are one of nature's most valuable resources, and their conservation is critical to the health of the entire planet (→ Keystone Species).

References

Drymon, James Marcus, Amanda E. Jargowsky, Evan G. Prasky, et al. 2024. "Depredation: An Old Conflict with the Sea." *Fish and Fisheries* 25: 806–810.

Guerra, Ana Sofia. 2019. "Wolves of the Sea: Managing Human-Wildlife Conflict in an Increasingly Tense Ocean." *Marine Policy* 99: 369–373.

Huveneers, Charlie, Craig Blount, Corey J.A. Bradshaw, et al. 2024. "Shifts in the Incidence of Shark Bites and Efficacy of Beach-Focused Mitigation in Australia." *Marine Pollution* Bulletin 298: 225855.

Martin, Carol. L., Belinda Curley, Kim Wolfenden, Marcel Green, and Natalie A. Moltschaniwskyj. 2022. "The Social Dimension to the New South Wales Shark

Management Strategy, 2015–2020, Australia: Lessons Learned." *Marine Policy* 141: 105079.

Mollier, Margaux, Camille Mazé, François Prioul, Paul Hamer, Malo Hosken, Sam McKechnie, and Paul Tixier. 2024. "Shark and Odontocete Depredation on the Catch of the Tuna Longline Fishery in New Caledonia (South Pacific Ocean)." *ICES Journal of Marine Science* 81 (3): 491–504.

NOAA Fisheries. 2022. "Celebrating 50 Years of the Marine Mammal Protection Act." www.fisheries.noaa.gov/feature-story/celebrating-50-years-marine-mammal-protection-act.

Noriega, Fernando, Vincent Lecours, and Luis Medrano-González. 2023. "Characterizing Spatio-Temporal Patterns of Coastal Navigation to Inform Mitigation Strategies for Conflict Between Ocean Users and the Protected Humpback Whale (*Megaptera novaeangliae*) in Bahía de Banderas, Mexico." *Regional Studies in Marine Science* 66: 103134.

Pacoureau, Nathan, Cassandra L. Rigby, Peter M. Kyne, et al. 2021. "Half a Century of Global Decline in Oceanic Sharks and Rays." *Nature* 589: 567–571.

Tixler, Paul, Mary-Anne Lea, Mark A. Hindell, Dirk Welsford, Camille Mazé, Sophie Gourguet, and John P.Y. Arnould. 2021. "When Large Marine Predators Feed on Fisheries Catches: Global Patterns of the Depredation Conflict and Directions for Coexistence." *Fish and Fisheries* 22: 31–53.

Conflict with Terrestrial Animals

Conflict with terrestrial animals describes events, actions, or threats that occur between people and wildlife that have an adverse effect (or the perception of a threat) on the other to their safety, health, or food (Nyhus 2016). The negative effects or actions of terrestrial species on human health and livelihoods have driven humans to persecute, extirpate, and cause severe range reduction of wolves in Asia, North America, and Europe, jaguars in the Americas, lions and wild dogs in Africa, and tigers in Asia (Nyhus 2016). Felids and canids are often in conflict with humans because they are large carnivores with vast home ranges. Approximately two dozen species of terrestrial carnivores prey on nine common species of livestock (Nyhus 2016). Historically, the human response to competition with carnivores over food resources has been to remove the competitors, reducing biodiversity and destroying ecosystems that are sustained by large apex predators (→ Habitat Loss) (→ Keystone Species) (→ Trophic Cascades). The removal of wolves from part of their range, for example, not only changes wolf pack structure and behavior but also increases populations of mesopredators such as coyotes who are controlled by top carnivores (→ Mesopredator Release) (Carter et al. 2019).

Human conflict with bears varies widely among eight species of bears and by geographic location, with brown and black bears the most often involved species (Krofel et al. 2020). Brown bears are the planet's most widely distributed mammals. They are generalist feeders and consume a variety of foods including livestock, crops, and even beehives (Nyhus 2016).

Andean, sun, sloth, and polar bears are infrequently involved in conflict with humans, and the giant panda has never been reported as part of a conflict (Krofel et al. 2020), except in zoo environments (Zhang et al. 2014). One of the most effective ways to mitigate conflict with bears is to limit availability of anthropogenic food particularly in residential areas and national parks (→ Coexistence). In addition, electric fences and rapid removal of livestock carcasses have reduced cattle depredation in South America. In the Arctic where their habitat has been damaged by climate change, hungry polar bears who approach human settlements are captured and either placed in a holding facility for at least 30 days or relocated outside of town (→ Habitat Loss) (→ Climate Change) (Krofel et al. 2020).

Unlike the contexts of all other human-carnivore conflicts, crocodilians (alligators, caimans, crocodiles, and gharial) are unique in their impact on both terrestrial and aquatic ecosystems (Cavalier et al. 2022). Crocodilians help maintain ecosystem balance as apex predators (→ Keystone Species). American alligators enhance wetland food-web heterogeneity by increasing nutrient availability, manipulating physical structure, and altering algal, plant, and animal communities (→ Trophic Cascades) (Strickland et al. 2023). But the conflict between humans and crocodilians is substantial. Although remote communities are at risk of fatal alligator attacks, most human-crocodilian reports are nuisance conflicts in "hot spots" of growing human populations and residential and public areas including golf courses, parks, and schools (Cavalier et al. 2022). In a study of four decades of alligator bites in Florida, Woodward et al. (2019) found that most incidents were attempts at feeding by people, and 37 percent entailed a single bite followed by an immediate release indicating that alligators were biting in defense or were unsure about their prey. Conflict with alligators can be reduced by selectively removing problem alligators from residential areas and water bodies and by employing strategies that attend to "humans' ability and willingness to more fully understand how to share space with these living dinosaurs" (→ Coexistence) (Cavalier et al. 2022, 391).

Conflicts with large vertebrate herbivores (e.g., elephants, swine, deer, hippopotami) are usually caused by their trampling, consuming, and otherwise damaging important vegetation, particularly elephants who bring damage to crops and vegetation in Asia and Africa (Nyhus 2016). Wild boars are also damaging to livestock and crops, destruction that cost the US economy $800 million in 2005 (Tampakis et al. 2023). Wild boars destroy plant variety, composition, and renewal, cause road accidents, injure domestic pets, and spread bacterial infections (Tampakis et al. 2023). Mitigation strategies include banning the feeding of wild boars and the use of electrical fencing in rural areas (→ Coexistence). Controlling the wild boar population needs to be approached holistically. The wild boar is another ecosystem engineer who controls parasitic species and contributes to forest regeneration and the growth of new vegetation by digging up surface soil in the search for food (→ Ecosystem Engineers) (Tampakis et al. 2023).

Humans also have conflict with wild birds, particularly those whose decreased populations have now rebounded with conservation efforts. Human-bird conflicts are linked to agricultural interests, forestry, hunting, fisheries, energy production, and public health. In an analysis of levels of human-bird conflict, McMahon et al. (2024) conclude that the lowest level of conflict includes nuisance caused by gulls in urban environments and bird damage to crops. The next level of conflict intensity relates to underlying conflicts, where past activities intensify or aggravate the current situation, such as the reintroduction of white-tailed eagles in Scotland and Ireland who were extirpated in the early twentieth century and are now impacting agricultural productivity and farmers' livelihoods. The highest level of conflict are deep-rooted, identity-based struggles, such as the conflict over conservation of the Spotted Owl in North America between those who value old-growth forests and those who favor lumbering (→ Endangered Species Act), and the conflict over the protection of Mute Swans who overfeed on plants that retain the ecosystem of chalk rivers in England. Human-wild bird conflicts generate strong emotions from multiple stakeholders. Mitigation strategies need to be interdisciplinary and capable of empowering ornithologists and bird conservationists to contribute to potential management approaches (→ Coexistence) (McMahon et al. 2024).

Climate change exacerbates human-wildlife conflict (→ Climate Change). A study of the influence of drought on reports of human-wildlife conflict found that prolonged decreases in precipitation are associated with increased overall conflict occurrences, particularly with carnivore species (Calhoun et al. 2025). Intensifying droughts driven by climate change are likely to lead to worsening conflicts with terrestrial animals and understanding how climate-driven conflict affects conservation is becoming more important in our changing world (Calhoun et al. 2025).

References

Calhoun, Kendall L., Justine A. Smith, Morgan W. Tingley, et al. 2025. "Human-wildlife Conflict Is Amplified During Periods of Drought." *Science Advances* 11: eadx0286.

Carter, Neil H., Jeremy T. Bruskotter, John Vucetich, et al. 2019. "Towards Human–Wildlife Coexistence through the Integration of Human and Natural Systems: The Case of Grey Wolves in the Rocky Mountains, USA." In *Human–Wildlife Interactions: Turning Conflict into Coexistence*, edited by Beatrice Frank, Jenny A. Glikman, and Silvio Marchini, 384–413. Cambridge University Press.

Cavalier, Rebecca, Elizabeth N. Pratt, Christopher Serenari, and Elena C. Rubino. 2022. "Human Dimensions of Crocodilians: A Review of the Drivers of Coexistence." *Human Dimensions of Wildlife* 27 (4): 380–396.

Krofel, Miha, Marcus Elfström, Hüseyin Ambari, et al. 2020. "Human–Bear Conflicts at the Beginning of the Twenty-First Century: Patterns, Determinants, and Mitigation Measures." In *Bears of the World: Ecology, Conservation and Management*, edited by Vincenzo Penteriani and Mario Melletti. Cambridge University Press. https://doi.org/10.1017/9781108692571.016.

McMahon, Barry J., Beatriz Arroyo, Nils Bunnefeld, Martina Carrete, Francis Daunt, and Juliette C. Young. 2024. "Birds and People: From Conflict to Coexistence." *International Journal of Avian Science* 166: 23–37.

Nyhus, Philip J. 2016. "Human-Wildlife Conflict and Coexistence." *Annual Review of Environment and Resources* 41: 143–174.

Strickland, Bradley A., Peter J. Flood, Jeffrey L. Kline, Frank J. Mazzotti, Michael R. Heithaus, and Joel C. Trexler. 2023. "An Apex Predator Engineers Wetland Food-Web Heterogeneity through Nutrient Enrichment and Habitat Modification." *Journal of Animal Ecology* 92 (7): 1287–1469.

Tampakis, Stilianos, Veronika Andrea, Thomas Panagopoulos, Paraskevi Karanikola, Rallou Gkarmiri, and Theodora Georgoula. 2023. "Managing the Conflict of Human–Wildlife Coexistence: A Community-Based Approach." *Land* 12, 832. https://doi.org/10.3390/land12040832.

Woodward, Allan R., Erin H. Leone, Harry J. Dutton, Jason E. Waller, and Lindsey Hord. 2019. "Characteristics of American Alligator Bites on People in Florida." *The Journal of Wildlife Management* 83 (6): 1437–1453.

Zhang, Peixun, Tianbing Wang, Jian Xiong, et al. 2014. "Three Cases Giant Panda Attack on Human at Beijing Zoo." *International Journal of Clinical and Experimental Medicine* 7 (11): 4515–4518. https://pmc.ncbi.nlm.nih.gov/articles/PMC4276236/.

Conservation and Green Criminology

Criminology is the study of crime, criminal behavior, crime prevention, law enforcement, and the criminal justice system. Conservation criminology applies insights from criminology to understand and prevent crime that has environmental consequences, especially crime that is destructive to ecosystems or harmful to other species (Gore 2017; McGarrell and Gibbs 2014). Conservation criminology is an interdisciplinary and applied approach to understanding programs and policies associated with global conservation risks. By integrating natural resources management, risk and decision sciences, and criminology, conservation criminology-based approaches ideally result in improved environmental resilience, biodiversity conservation, and secure human livelihoods (→ Biodiversity). A substantial focus has been on crimes that lead to poaching and illegal trafficking of animals and animal products (→ Wildlife Crime) (St. John et al. 2025). Transnational environmental crime is of such a large magnitude globally that it constitutes a series threat to sustainability (Gore et al. 2019). In the US alone, 8-9000 game wardens, law enforcement rangers, and other officers focus on environmental crime.

Green criminology is similar to conservation criminology in its focus on activities that damage the environment that are not directly related to traditional conservation concerns. Much green criminology scholarship focuses on crimes that are committed by corporations and governments. This expanded scope would consider actions such as illegal dumping of toxic waste or falsifying records of environmental compliance. But since any harm to the environment is a challenge to sustainability and most have effects that ripple through ecosystems, the boundaries between green and conservation criminology are not rigid. Recent work extends these perspectives to include critical examination of actions that damage the environment even if they are not illegal and engages with theories of political economy (Beirne and South 2007; Long et al. 2024; van Uhm 2024). There is also a growing effort to link green and conservation criminology with environmental justice since the harms to the environment being examined often are highly inequitable in their impacts (→ Environmental Justice) (van Uhm and Siegel 2025). Some scholars argue that green and conservation criminology should include consideration of cruelty to non-human animals as part of its focus (Taylor and Fitzgerald 2018).

References

Beirne, Piers, and Nigel South, eds. 2007. *Issues in Green Criminology: Confronting Harms against Environments, Humanity and Other Animals*. Willan.

Gore, Meredith L., ed. 2017. *Conservation Criminology*. Wiley/Blackwell.

Gore, Meredith L., Patrick Braszak, James Brown, et al. 2019. "Transnational Environmental Crime Threatens Sustainable Development." *Nature Sustainability* 2 (9): 784–786. https://doi.org/10.1038/s41893-019-0363-6.

Long, Michael A., Paul B. Stretesky, and Michael J. Lynch. 2024. "Green Criminology." In *Encyclopedia of Technological Hazards and Disasters in the Social Sciences*, edited by Duane A. Gill, Liesel A. Ritchie, and Nnenia M. Campbell, 295–301. Edward Elgar Publishing.

McGarrell, Edmund F., and Carole Gibbs. 2014. "Conservation Criminology, Environmental Crime, and Risk: An Application to Climate Change." In *Oxford Handbooks Topics in Criminology and Criminal Justice*. Oxford Academic, 2 June 2014. https://doi.org/10.1093/oxfordhb/9780199935383.013.54.

St. John, Freya A.V., Leejiah Dorward, Harriet Ibbett, and Martina Feilzer. 2025. "Using Lessons from Criminal Justice Research to Improve Conservation Law Enforcement Research and Practice." *Conservation Biology*, July 3, e70094. https://doi.org/10.1111/cobi.70094.

Taylor, Nik, and Amy Fitzgerald. 2018. "Understanding Animal (Ab)use: Green Criminological Contributions, Missed Opportunities and a Way Forward." *Theoretical Criminology* 22 (3): 402–425. https://doi.org/10.1177/1362480618787173.

van Uhm, Daan. 2024. "Green Criminology and Biodiversity Loss: Crimes and Harms against Flora and Fauna." In *Oxford Research Encyclopedia of Criminology and Criminal Justice*, edited by Henry N. Pontell. Oxford University Press. https://doi.org/10.1093/acrefore/9780190264079.013.763.

van Uhm, Daan, and Dina Siegel, eds. 2025. *Global Green Crime and Ecojustice. Palgrave Studies in Green Criminology*. Palgrave Macmillan. https://doi.org/10.1007/978-3-031-91557-4

Conservation Biology

Conservation biology "addresses the biology of species, communities, and ecosystems that are perturbed, either directly or indirectly, by human activities or other agents ... (its) goal is to provide principles and tools for preserving biological diversity" (Soulé 1985, 727). It is a crisis-oriented discipline, with scientists asked to advise on a variety of ecological and health problems, such as the introduction of exotic species and artificially produced strains of existing organisms, the sizes and sitings of national parks, the minimum conditions for viable populations of species, the management of wildlands and refuges, and the ecological effects of development (Soulé 1985). One of the distinguishing characteristics of conservation biology is its holistic, multidisciplinary approach to problems that engage a variety of disciplines such as ecology, population biology, natural resource fields, genetics, social sciences, ecophilosophy, and historical biogeography (→ Ecology) (→ Biotechnology) (→ Island Biogeography) (Soulé 1985). In addition, Lyman (2006) argues for the use of paleoecological knowledge in conservation biology because conservation biologists often need an ecological baseline or benchmark to serve as a goal for conservation activities in an area, such as the presence or absence of a species or what species have been introduced by humans (→ Applied Zooarchaeology).

There are seven principles for the conservation of wild living resources often invoked in conservation biology (Mangel et al. 1996): 1) Maintenance of healthy populations of wild living resources in perpetuity is inconsistent with unlimited growth of human consumption of and demand for those resources; 2) The goal of conservation should be to secure present and future options by maintaining biological diversity at genetic, species, population, and ecosystem levels; as a general rule neither the resource nor other components of the ecosystem should be perturbed beyond natural boundaries of variation; 3) Assessment of the possible ecological and sociological effects of resource use should precede both proposed use and proposed restriction or expansion of ongoing use of a resource; 4) Regulation of the use of living resources must be based on understanding the structure and dynamics of the ecosystem of which the resource is a part and must take into account the ecological and sociological influences that directly and indirectly affect resource use; 5) The full range of knowledge and skills from the natural and social sciences must be brought to bear on conservation problems; 6) Effective conservation requires understanding and taking account of the motives, interests, and values of all users and stakeholders, but not by simply averaging their positions; and 7) Effective conservation requires communication that is interactive, reciprocal, and continuous.

References

Lyman, R. Lee. 2006. "Paleozoology in the Service of Conservation Biology." *Evolutionary Anthropology* 15: 11–19.

Mangel, Marc, Lee M. Talbot, Gary K. Meffe, et al. 1996. "Principles for the Conservation of Wild Living Resources." *Ecological Applications* 6 (2): 338–362.

Soulé, Michael E. 1985. "What Is Conservation Biology?" *BioScience* 35 (11): 727–734.

Conservation Finance for Animals and Ecosystems

Conservation finance for animals and ecosystems is an activity that raises and manages capital to be used to conserve biodiversity, such as funds for acquiring habitat, education, research, and other actions (Cosma et al. 2023). Proposals for conservation finance usually focus not just on biodiversity conservation but also on improving human well-being (Le et al. 2024). Humans are in poverty in many areas where animal species are threatened so moving toward sustainability requires attention to humans as well as other species and the environment.

At present the funds to support these activities come primarily from governments and from non-profit organizations, but most studies argue that the amount of funding required to protect biodiversity and move toward sustainability will also require private investments. If so, conservation funding mechanisms must be found to align with the needs of investors. Prominent attempts to do this include payments to governments or landowners under approaches called Reducing Emissions from Deforestation and Forest Degradation (REDD) and Payment for Ecosystem Services (PES) (Watson et al. 2024; Le et al. 2024). REDD programs provide funds based on the ability of forests to mitigate greenhouse gas emissions, while PES programs are broader and take consideration of other ecosystem services provided by intact ecosystems, but both reward the preservation of ecosystems (→ Ecosystem Services). The latest approaches to REDD, called REDD+, also take account of biodiversity and other benefits of forest ecosystems, the need for improvements in human wellbeing, and overall sustainability goals.

While conservation finance is often considered essential to sustainability, there are concerns with current efforts. Funds tend to be targeted at charismatic species, with many other species neglected (Hughes et al. 2025). For example, more funds are allocated to elephant programs than to all 12,000 reptile species combined, and very little funding is allocated to invertebrates. Some argue that conservation finance needs to be more fully integrated with broader efforts to move toward sustainability, such as providing basic income support for the poor (Fletcher et al. 2025). Some also critique conservation finance as re-enforcing growth dynamics and unequal distribution of power (Dehm 2021). Overall, there is concern that while conservation finance mechanisms have helped protect species and move toward sustainability, there continue to be problems with both effective implementation and with increasing the scale of conservation finance to match the scope of global problems (Morita and Matsumoto 2023).

References

Cosma, Simona, Giuseppe Rimo, and Stefano Cosma. 2023. "Conservation Finance: What Are We Not Doing? A Review and Research Agenda." *Journal of Environmental Management* 336 (June): 117649. https://doi.org/10.1016/j.jenvman.2023.117649.

Dehm, Julia. 2021. *Reconsidering REDD+: Authority, Power, and Law in the Green Economy*. Cambridge Studies on Environment, Energy and Natural Resources. Cambridge University Press.

Fletcher, Robert, Emiel de Lange, Isabel Felandro, et al. 2025. "Cash for Conservation? Integrating Basic Income Support into Biodiversity and Climate Finance." *Current Opinion in Environmental Sustainability* 75 (August): 101554. https://doi.org/10.1016/j.cosust.2025.101554.

Hughes, Alice C., Stefano Cannicci, Bayden Russell, Gray A. Williams, and Benoit Guénard. 2025. "Toward a More Effective Funding Model for Conservation." *BioScience* 76 (1): 10–12. https://doi.org/10.1093/biosci/biaf124.

Le, Tuyet-Anh T., Kelly Vodden, Jianghua Wu, Ryan Bullock, and Gabriela Sabau. 2024. "Benefits and Risks from Payments for Ecosystem Services Programs across the Globe." *Frontiers in Environmental Science* 12 (July): 1419821. https://doi.org/10.3389/fenvs.2024.1419821.

Morita, Kanako, and Ken'ichi Matsumoto. 2023. "Challenges and Lessons Learned for REDD+ Finance and Its Governance." *Carbon Balance and Management* 18 (1): 8. https://doi.org/10.1186/s13021-023-00228-y.

Watson, Charlene, Liane Schalatek, and Aurélien Evéquoz. 2024. *Climate Finance Thematic Briefing: REDD+ Finance*. Heinrich Böll Stiftung. https://climatefundsupdate.org/wp-content/uploads/2024/04/CFF5-2024-ENG-REDD-DIGITAL.pdf.

Conservation Psychology

Conservation psychology is an interdisciplinary field that examines the psychological factors that shape how humans understand, value, and respond to the natural world. Emerging in the early 2000s, conservation psychology integrates insights from cognitive psychology, social psychology, environmental studies, sociology, and human-animal studies to explain why people engage in behaviors that either support or undermine environmental protection (Saunders 2003). Conservation psychology is often conceptualized as either a subfield of environmental psychology or a closely related discipline, distinguished by its primary focus on human relationships with wildlife and ecosystems rather than on broader categories of environmentally consequential behavior. Central to the field is the idea that environmental issues such as biodiversity loss, habitat degradation, and climate change are not only ecological problems, but also cognitive and behavioral ones (→ Biodiversity) (→ Climate Change) (→ Habitat Loss). Conservation psychology focuses on attitudes, motivations, emotions, and social contexts that influence individual and collective action toward protecting ecosystems and the species who inhabit them (Clayton and Myers 2015).

A clear example of conservation psychology's relevance to non-human animals can be seen in research on public attitudes toward wildlife (→ Wilderness Animals). Studies have shown that people's emotional responses, such as empathy, fear, or fascination, affect support for conservation initiatives aimed at species ranging from elephants to amphibians (→ Empathy) (Schultz 2011). For instance, work on carnivore conservation demonstrates that perceptions of risk, beliefs, and trust in management agencies heavily influence willingness to coexist with wolves and large cats (→ Coexistence) (Carlson et al. 2020). Zoos and aquariums also draw heavily on conservation psychology by attempting to design educational programs and exhibits that foster emotional connection and pro-environmental behavior among visitors (→ Zoos and Aquariums) (Clayton et al. 2009). These programs often rely on evidence that meaningful encounters with animals or animal "stand-ins" such as statues can motivate long-term concern for wildlife and habitat protection (→ Animal Ambassadors) (Wilson et al. 2022).

Conservation psychology contributes to sustainability by linking human behavior, non-human animal well-being, and ecological health. Because sustainable futures require both reducing human impacts to ecosystems and promoting coexistence with other species, understanding the psychological drivers of conservation action becomes essential. Research shows that fostering empathy for animals, strengthening environmental identities, and creating opportunities for behavioral change can enhance public support for conservation policies and reduce activities that harm wildlife (Clayton and Myers 2015). In this way, conservation psychology helps bridge the gap between scientific knowledge and practical action, offering tools that encourage individuals and communities to adopt sustainable practices that benefit both humans and non-human animals.

References

Carlson, Shelby C., Alia M. Dietsch, Kristina M. Slagle, and Jeremy T. Bruskotter. 2020. "The VIPs of Wolf Conservation: How Values, Identity, and Place Shape Attitudes toward Wolves in the United States." *Frontiers in Ecology and Evolution* 8: 6.

Clayton, Susan, and Gene Myers. 2015. *Conservation Psychology: Understanding and Promoting Human Care for Nature*. Wiley.

Clayton, Susan, John Fraser, and Carol D. Saunders. 2009. "Zoo Experiences: Conversations, Connections, and Concern for Animals." *Zoo Biology* 28 (5): 377–397.

Saunders, Carol D. 2003. "The Emerging Field of Conservation Psychology." *Human Ecology Review*, 137–149.

Schultz, P. Wesley. 2011. "Conservation Means Behavior." *Conservation Biology* 25 (6): 1080–1083.

Wilson, Jennifer K., Nichole L. Nageotte, and Nette Pletcher. 2022. "Evoking Empathy — An Overlooked Role of Animal Statues in Zoos." *Zoo Biology* 41 (5): 418–424.

Convention on Biological Diversity

The Convention on Biological Diversity (CBD) is an international agreement in which nations commit to protect biodiversity in fair and equitable ways (→ Biodiversity) (https://www.cbd.int/). It was initiated at the Rio Earth Summit in 1992, at the same time as the United Nations Framework Convention on Climate Change. Together the agreements address two major aspects of global environmental change (→ Global Environmental Change).

In 2010, the CBD set 20 ambitious goals with the Aichi Targets (Hughes 2023), but by 2020, it was clear that none of these targets had been met. That led to the 2022 Kunming-Montreal Global Biodiversity Framework (KM-GBF) that set the "30 by 30" goals where nations pledged to set aside 30 percent of the Earth's land area and 30 percent of the oceans for biodiversity protection by 2030. Most nations have ratified this target and there appears to be substantial public support for it (Michaelsen et al. 2025). Progress has been made in increasing protected areas, with an estimated 18 percent of land area and 8 percent of oceans protected (United Nations Environment Programme and International Union for the Conservation of Nature 2024). Only about 3 percent of the ocean area is considered "fully" or "highly" protected. Clearly much more needs to be done if the goals are to be achieved. While the 30 by 30 goals will provide protection for many animal species, there is growing concern that migratory animals may need additional consideration as they move across ecosystems (Davis et al. 2024).

The Intergovernmental Science-Policy Platform on Biodiversity and Ecosystem Services (IPBES) was established by agreement of 94 government in 2012. It is an ongoing scientific assessment that is intended to provide the scientific analyses that CBD and the participating nations need to make sound decisions. Currently, 150 nations participate, and IPBES generates scientific assessment reports on many topics related to biodiversity. For example, recent reports highlight the importance of business engagement in the protection of biodiversity (Jones et al. 2026), the transformation needed to achieve the biodiversity goals (O'Brien et al. 2025), and the connections between biodiversity and several other key aspects of sustainability such as water, food, and health (→ Sustainable Development Goals) (Harrison et al. 2024). IPBES is one of an ongoing series of international scientific assessments intended to summarize science for decision making (Bennett et al. 2026).

References

Bennett, Elena, Walt Reid, Stephen R. Carpenter, et al. 2026. "The Future of Global Environmental Assessments: 20 Years after the Millennium Ecosystem Assessment." *Current Opinion in Environmental Sustainability* 79: 101618.

Davis, Frances, Andrew Szopa-Comley, Sarah Rouse, and Aude Caromel et al. 2024. *State of the World's Migratory Species*. UNEP-WCMC Cambridge. www.cms.int/sites/default/files/publication/State%20of%20the%20Worlds%20Migratory%20Species%20report_E.pdf.

Harrison, Paula, Pamela McElwee, and Tiff van Huysen. 2024. *IPBES Nexus Assessment: Full Report*. https://doi.org/10.5281/ZENODO.13850054.

Hughes, Alice C. 2023. "The Post-2020 Global Biodiversity Framework: How Did We Get Here, and Where Do We Go Next?" *Integrative Conservation* 2 (1): 1–9. https://doi.org/10.1002/inc3.16.

Jones, Matt, Stephen Polasky, Ximena Rueda, et al. 2026. *IPBES Business and Biodiversity Assessment: Summary for Policymakers*. Zenodo. https://doi.org/10.5281/ZENODO.18538597.

Michaelsen, Patrik, Aksel Sundström, and Sverker C. Jagers. 2025. "Mass Support for Conserving 30% of the Earth by 2030: Experimental Evidence from Five Continents." *Proceedings of the National Academy of Sciences* 122 (35): e2503355122. https://doi.org/10.1073/pnas.2503355122.

O'Brien, Karen, Lucas Garibaldi, and Arun Agrawal. 2025. *IPBES Transformative Change Assessment: Full Report*. Version v2. Zenodo. https://doi.org/10.5281/ZENODO.11382215.

United Nations Environment Programme, and International Union for the Conservation of Nature. 2024. *Protected Planet Report 2024*. Cambridge, UK and Gland, Switzerland. https://digitalreport.protectedplanet.net.

Convention on International Trade in Endangered Species (CITES)

The global trade in wildlife is a major driver of extinction risk for many species (→ Wildlife Trade). National governments use domestic laws, such as the US Endangered Species Act, to regulate wildlife trade (→ Endangered Species Act). The global community has responded with the Convention on International Trade in Endangered Species of Wild Flora and Fauna (CITES) (CITES 2025), an international treaty that went into effect in 1975 with 185 countries currently participating.

Over 40,000 species are covered by CITES. Species can be on one of three lists. Appendix I consist of species at risk of extinction and are given the highest level of protection. No commercial trade is allowed for these species, and other transfers of animals or plants, such as for research on conservation purposes, are strictly regulated. Appendix II are species where extensive trade might lead to extinction risk. Permits are required for any trade in these species. Appendix III are species listed as at risk by at least one country that participates in the treaty. The International Union for the Conservation of Nature (IUCN), a non-governmental organization founded in 1948, supported the creation of CITES (IUCN 2025). IUCN's Red List of Threatened Species is a crucial input into IUCN's decision about which species are listed in Appendices I, II, and III.

CITES is generally seen as successful in moderating wildlife trade but faces many challenges (Clark 2022; Challender et al. 2025). Illegal trade, including activities facilitated by online markets, is difficult to control (→ Wildlife Crime) (→ Conservation and Green Criminology). CITES is limited in scope since it focuses on international trade, not on other stresses such as climate change and habitat loss that put many species at risk of extinction (→ Climate Change) (→ Habitat Loss). CITES and the IUCN continue to evolve to better address threats of extinction and of overall sustainability (→ Extinction).

References

Challender, Daniel W.S., Michael 'T Sas-Rolfes, Steven Broad, and E.J. Milner-Gulland. 2025. "A Theory of Change to Improve Conservation Outcomes through CITES." *Frontiers in Ecology and Evolution* 13 (February): 1425267. https://doi.org/10.3389/fevo.2025.1425267.

Clark, Jeremiah D. 2022. "A Review of CITES's Impact and Suggestions for Incremental Improvement." *Temple International & Comparative Law Journal* 37: 127. https://heinonline.org/HOL/Page?handle=hein.journals/tclj37&id=131&div=&collection=.

CITES. 2025. Convention on International Trade in Endangered Species of Flora and Fauna. https://cites.org/eng.

IUCN. 2025. "Seventy-Five Years of Experience | International Union for the Conservation of Nature." https://iucn.org/about-iucn/history.

Coupled Human and Natural Systems

Coupled Human and Natural Systems (CHANS) is a conceptual framework that emphasizes the interconnections between human societies and ecosystems (Liu et al. 2007). It views human societies as embedded in ecosystems and as impacting ecosystems and holds that integrated analysis is essential for moving toward sustainability. All ecosystems on Earth and the animals in them are influenced by human actions; even the ecosystems surrounding thermal vents in the deepest part of the ocean are vulnerable (Menini et al. 2023). Some of these influences come from direct human intervention, others from systemic global environmental change (→ Global Environmental Change). In turn, human societies are dependent on the myriad benefits provided by ecosystems (→ Ecosystem Services). The CHANS perspective encourages examining these links rather than considering human or ecological systems in isolation.

CHANS is closely linked to and builds upon other perspectives emphasizing the connection between human and ecological systems, such as Ostrom's Social-Ecological Systems (SES) and the field of human ecology (→ Ecology) (Ostrom 2009). These perspectives are all efforts to encourage an approach in science that parallels ecocentric thinking in ethics – avoiding analyses that ignore non-human species or ecosystems (→ Ecocentrism). Non-human animals have always been central to CHANS analysis. The perspective emerged from studies of interactions of humans, pandas, and other species in the Wolong Nature Reserve for Giant Pandas (Liu et al. 2016). The way farmers manage their land is crucial to the availability of habitat for pandas. With the help of supportive government policies, humans living in the reserve shifted their activities to be more "panda friendly" (→ Coexistence). Residents have engaged with conservation by reporting panda sitings to researchers and participating in panda tourism (→ Wildlife Tourism). Sustainability of the panda habitat is also enhanced by China's Grain to Green Program and the Natural Forest Conservation Program that have curbed deforestation, social erosion, illegal logging, and harvesting fuelwood from forests.

References

Liu, Jianguo, Thomas Dietz, Stephen R. Carpenter, et al. 2007. "Complexity of Coupled Human and Natural Systems." *Science* 317 (5844): 1513–1516.

Liu, Jianguo, Vanessa Hull, Wu Yang, et al. 2016. *Pandas and People: Coupling Human and Natural Systems for Sustainability*. Oxford University Press.
Menini, Elisabetta, Helena Calado, Roberto Danovaro, Elisabetta Manea, and Patrick N. Halpin. 2023. "Towards a Global Strategy for the Conservation of Deep-Sea Active Hydrothermal Vents." *Npj Ocean Sustainability* 2 (1): 22. https://doi.org/10.1038/s44183-023-00029-3.
Ostrom, Elinor. 2009. "A General Framework for Analyzing Sustainability of Social-Ecological Systems." *Science* 325: 419–422.

Critical Animal Studies

Critical animal studies (CAS) is an interdisciplinary field that examines human-animal relationships through a lens of justice, power, and liberation. Emerging in the early 2000s from animal advocacy, environmental ethics, and sociology, CAS distinguishes itself from more traditional animal studies by its normative commitment to challenging systems of exploitation and inequality (Best et al. 2007). Scholars in this field draw from cultural studies, geography philosophy, political science, and sociology, to interrogate how animals are embedded in structures such as capitalism, colonialism, and patriarchy (→ Colonialism). Rather than treating animals solely as objects of study, CAS considers questions of ethics, agency, and resistance, while aligning scholarship with activism (→ Agency) (→ Advocacy). This orientation reflects a normative shift within the social sciences toward engaged and reflexive research that seeks not only to understand the world, but to transform it (Twine and Taylor 2014; Nibert 2013).

In practice, CAS attempts to engage work that highlights the intersections between animal exploitation and other forms of social inequality. For example, research within CAS has examined industrial animal agriculture as a site where environmental degradation, labor exploitation, and animal suffering converge, illustrating how systems of production simultaneously harm humans and nonhuman animals (→ Livestock Production) (Nibert 2013; Brookes and Chałupnik 2025; Twine 2010). CAS scholars also have detailed how cultural representations of animals reinforce hierarchies of worth, shaping public attitudes toward different species and legitimizing unequal treatment (Twine and Taylor 2014). They have also informed critiques of institutions such as zoos, laboratories, and industrial farms, encouraging greater attention to issues of power, voice, and representation (see e.g., Hooper 2023; Malamud 2015). Importantly, the field is closely connected to activist movements, with many scholars advocating for policy changes, alternative food systems, and expanded legal protections for animals (Best et al. 2007; Best 2009, 2014; Tomasello et al. 2025). CAS has been critiqued for its explicit normative commitments, theoretical eclecticism, radical ethical claims, exclusion or minimization of perspectives from Indigenous, disability, and other marginalized communities, and limits to practical applicability (→ Indigenous and Decolonial Perspectives) (→ Disability) (Shapiro 2020; Belcourt 2020; Jenkins et al. 2020).

From a sustainability perspective, CAS offers an important framework for rethinking human relationships with other species in the context of ecological crisis. Rather than focusing solely on conservation outcomes, CAS emphasizes the ethical dimensions of sustainability, arguing that long-term ecological resilience is inseparable from questions of use, justice, and equity (Twine 2010). Recent work suggests a convergence between critical sustainability studies and critical animal studies (→ Critical Sustainability Studies) (Bergmann 2020). This approach encourages scholars and practitioners to consider how dominant systems of consumption, particularly those involving animal use, contribute to climate change, biodiversity loss, and resource depletion (→ Climate Change) (→ Biodiversity). As environmental challenges intensify, the field may provide critical tools for envisioning more just and sustainable multispecies futures (→ Coexistence) (→ Multispecies Futures).

References

Belcourt, Billy-Ray. 2020. "An Indigenous Critique of Critical Animal Studies." In *Colonialism and Animality*, edited by Kelly Struthers Montford, and Chloë Taylor. Routledge.

Bergmann, Iris M. 2020. "The Intersection of Animals and Global Sustainability — A Critical Studies Terrain for Better Policies?" *Proceedings of the 1st International Electronic Conference on Animals—Global Sustainability and Animals: Science, Ethics and Policy* 73 (1): 12.

Best, Steve, Anthony J. Nocella, Richard Kahn, Carol Gigliotti, and Lisa Kemmerer. 2007. "Introducing Critical Animal Studies." *Journal for Critical Animal Studies* 5 (1): 4–5.

Best, Steven. 2009. "The Rise of Critical Animal Studies: Putting Theory into Action and Animal Liberation into Higher Education." *Journal for Critical Animal Studies* 7 (1): 9–52.

Best, Steven. 2014. *The Politics of Total Liberation: Revolution for the 21st Century*. Springer.

Brookes, Gavin, and Małgorzata Chałupnik. 2025. "Hard to Swallow? A Critical Animal Studies Perspective on the Discursive Recontextualisation of the Reality of Dairy Farming." *Discourse & Communication* 19 (4): 613–638.

Hooper, Jes. 2023. "Thinking with Civets: The Role of Zoos in the Decolonisation of Animal Tourism." *Animals* 13 (11): 1739.

Jenkins, Stephanie, Kelly Struthers Montford, and Chloë Taylor. 2020. *Disability and Animality: Crip Perspectives in Critical Animal Studies*. Routledge.

Malamud, Randy. 2015. "The Problem with Zoos." In *The Oxford Handbook of Animal Studies*, edited by Linda Kalof. Oxford University Press.

Nibert, David. 2013. *Animal Oppression and Human Violence: Domesecration, Capitalism, and Global Conflict*. Columbia University Press.

Shapiro, Kenneth. 2020. "Human-Animal Studies: Remembering the Past, Celebrating the Present, Troubling the Future." *Society & Animals* 28 (7): 797–833.

Tomasello, Sarah, Erin Jones, Mark Suchyta, and Nathan Poirier. 2025. *Exploring Topics in Non/Human Coexistence: Passion, Praxis and Presence*. Lantern Publishing.

Twine, Richard. 2010. *Animals as Biotechnology: Ethics, Sustainability and Critical Animal Studies*. Routledge.
Twine, Richard, and Nik Taylor. 2014. *The Rise of Critical Animal Studies from the Margins to the Centre*. Routledge.

Critical Anthropomorphism

Critical anthropomorphism is an approach in animal behavior and animal studies that uses informed, scientifically grounded comparisons to interpret animals' experiences and behaviors while avoiding unreflective projection of human traits. Rather than assuming animals think or feel as humans do, it treats anthropomorphism (the attribution of human characteristics or behavior to animals or objects) as a cautious heuristic rooted in biology. It draws on ideas from ethology, neuroscience, comparative psychology, and evolution theory to generate testable questions about animal cognition and motivation (Burghardt 2014).

Debates over critical anthropomorphism focus on both its usefulness and its risks. Supporters argue that excluding anthropomorphic reasoning altogether can obscure meaningful behavioral cues), such as grief in elephants or play in corvids (de Waal 2016; Plotnik and de Waal 2014). Critics counter that even disciplined anthropomorphism may blur species-specific differences or lead to interpretations shaped more by human expectations than by evidence, particularly in settings where welfare decisions rely on accurate assessments of animal needs (→ Welfare) (Sherman et al. 2025; Horowitz and Bekoff 2007). Some scholars extend the framework through concepts such as anthro-interpretivism, which adapts interpretive anthropology to include non-human animals, encouraging empathetic approaches like perspective taking that situate animals as individuals embedded in social collectives (→ Empathy) (Taylor 2011). Critical anthropomorphism complements compassionate conservation by using empirically informed interpretations of animal experience to guide management decisions that minimize harm while respecting individual animals as moral subjects (→ Compassionate Conservation). Others emphasize strategic anthropomorphism, in which educators and conservationists deliberately use human-centered language and tools, such as animal photo portraiture, to evoke empathy while trying to remain faithful to presenting biologically relevant species-typical behavior (Whitley et al. 2025).

Critical anthropomorphism has direct implications for sustainability because perceptions of animals shape environmental ethics, policy support, and conservation behavior. Research shows that anthropomorphic framing can increase feelings of connection to nature, strengthen concern for environmental problems, and counter values that prioritize human dominance over ecosystems (→ Values) (Young et al. 2018). Some ethical theories, such as the capabilities approach and more political arguments for including consideration of non-humans in practice engage insights much like those of critical anthropomorphism to imagine the capabilities, wants and needs of other species (→ Capabilities Ethics) (→ Ecodemocracy). When grounded in science, it can help the public understand animals as sentient participants in ecological systems, promoting habitat protection, coexistence strategies, and welfare-oriented conservation (→ Sentience) (→ Agency) (→ Coexistence). At the same time, responsible use of anthropomorphism guards against misconceptions that can obscure ecological realities or privilege charismatic species. By balancing empathy with scientific restraint, critical anthropomorphism contributes to sustainability efforts that integrate ecological knowledge, ethical responsibility, and respect for non-human life.

References

Burghardt, Gordon M. 2014. "Cognitive Ethology and Critical Anthropomorphism: A Snake with Two Heads and Hognose Snakes That Play Dead." In *Cognitive Ethology: The Minds of Other Animals*, edited by Carolyn A. Ristau. Psychology Press.

de Waal, Frans. 2016. *Are We Smart Enough to Know How Smart Animals Are?* W.W. Norton & Company.

Horowitz, Alexandra C., and Marc Bekoff. 2007. "Naturalizing Anthropomorphism: Behavioral Prompts to Our Humanizing of Animals." *Anthrozoös* 20 (1): 23–35.

Plotnik, Joshua M., and Frans B.M. de Waal. 2014. "Asian Elephants (Elephas maximus) Reassure Others in Distress." *PeerJ* 2: e278.

Sherman, Wilson C., Ashley J.A. Terry, and Alison W. Bowers. 2025. "Identifying Indicators, Empathy, and Anthropomorphism in Zoo Visitors' Perceptions of Animal Wellbeing through Qualitative Interviews." *Curator: The Museum Journal* 68 (1): 141–161.

Taylor, Nik. 2011. "Anthropomorphism and the Animal Subject." In *Anthropocentrism: Humans, Animals, Environments*, edited by Rob Boddice, 265–279. Brill.

Whitley, Cameron T., Linda Kalof, L.C. Urquhart, et al. 2025. "Photo Portraiture Enhances Empathy for Birds with Potential Benefits for Conservation and Sustainability." *Sustainability* 17 (19): 8833.

Young, Ashley, Kathayoon A. Khalil, and Jim Wharton. 2018. "Empathy for Animals: A Review of the Existing Literature." *Curator: The Museum Journal* 61 (2): 327–343.

Critical Sustainability Studies

Critical sustainability studies (CSS) is an interdisciplinary field that interrogates sustainability through the lens of power, inequality, and political economy. It emerged out of critical theory, environmental justice, and radical extensions of sustainability science in the late 1990s and early 2000s and "involves critiques of capitalist constructions of sustainability, utilizing research and practice that explores how in a complex, dynamic, and power-laden world, people can develop societal pathways to sustainability that embrace ecological integrity and social justice" (Rose and Cachelin 2018, 522). A collection of scholars including Miriam Greenberg (2013), Felipe Ferreira (2017), Jeff Rose and Adrienne Cachelin (2018) began theorizing CSS frameworks from collective dissatisfaction around existing sustainability studies that often focus on growth and efficiency in capitalist systems. CSS emphasizes that sustainability is fundamentally about contested relationships between society and nature, shaped by unequal power relations and competing values (Fisher et al. 2022). Ongoing work in this area asserts that CSS should be grounded in critical realism and historical materialism, arguing that these traditions offer robust tools for analyzing the socioecological dynamics underpinning environmental challenges, and that such an approach could bridge the natural and social sciences while highlighting the structural conditions and constraints that encourage sustainability transformation (Longo et al. 2025). A similar approach has been proposed for animal studies (York and Longo 2017). Ultimately, rather than asking how to make current systems more sustainable, CSS asks whether those systems are the root of unsustainable practices and outcomes and theorizes about what should be done.

A useful example of CSS in connection with human-animal relationships can be found in debates over methane emissions from industrial animal agriculture (→ Livestock Production). A traditional sustainability approach would focus on improving efficiency by modifying systems to reduce methane emissions. By contrast, a CSS approach would interrogate the broader socio-political system that normalizes large-scale animal exploitation, where livestock production is recognized as embedded in global capitalism, relies on uneven labor conditions, and disproportionately impacts marginalized communities of humans and non-human animals, while subjecting non-human animals to confinement and death (→ Metabolic Rift Theory) (Bergmann 2021). From a CSS perspective, the issue is not simply how to make animal agriculture "greener," but how dominant food systems and individual choices reproduce ecological degradation and multispecies injustice (→ Multispecies Futures). This framing could lead to alternative practices such as plant-based transitions, agroecology, or multispecies justice frameworks. CSS encourages a shift away from technical fixes to structural transformation, linking environmental sustainability with ethical considerations in alignment with other fields such as critical animal studies and environmental justice (→ Values) (→ Critical Animal Studies) (→ Environmental Justice).

CSS offers a generative framework for rethinking the place of animals in sustainability discourse. It pushes a broader recognition of animals not just as resources or providers of ecosystem services, but as subjects embedded in power-driven social relationships. It invites more explicit engagement with intersectionality, examining how speciesism intersects with race, class, colonialism, and other axes of inequality (→ Intersectional Environmentalism) (→ Speciesism) (→ Colonialism) (Pellow 2025; Celermajer et al. 2022). CSS suggests that sustainable futures will require more than technological innovation but will demand a reconfiguration of human-animal relations grounded in justice, care, and coexistence (→ Coexistence). While CSS analyses usually do not offer a blueprint for sustainability, they encourage scholars to engage ethics and analyses of power to complement perspectives grounded in ecology and other traditional disciplines in examining sustainability.

References

Bergmann, Iris M. 2021. "The Intersection of Animals and Global Sustainability — A Critical Studies Terrain for Better Policies?" *Proceedings of the 1st International Electronic Conference on Animals—Global Sustainability and Animals: Science, Ethics and Policy* 73 (1): 12.

Celermajer, Danielle, David Schlosberg, Lauren Rickards, et al. 2022. "Multispecies Justice: Theories, Challenges, and a Research Agenda for Environmental Politics." *Trajectories in Environmental Politics* 30 (1-2): 119–140.

Ferreira, Felipe. 2017. "Critical Sustainability Studies: A Holistic and Visionary Conception of Socio-Ecological Conscientization." *The Journal of Sustainability Education* 13 (March). www.susted.com/wordpress/content/critical-sustainability-studies-a-holistic-and-visionary-conception-of-socio-ecological-conscientization_2017_04/.

Fisher, Eleanor, Eduardo Brondizio, and Emily Boyd. 2022. "Critical Social Science Perspectives on Transformations to Sustainability." *Current Opinion in Environmental Sustainability* 55: 101160.

Greenberg, Miriam. 2013. "What on Earth Is Sustainable? Toward Critical Sustainability Studies." *Boom: A Journal of California* 3 (4): 54–66.

Longo, Stefano B., Ellinor Isgren, and Michael Carolan. 2025. "Critical Sustainability Science: Advancing Sustainability Transformations." *Sustainability Science* 20 (5): 1903–1917.

Pellow, David Naguib. 2025. *What Is Critical Environmental Justice?* Wiley.

Rose, Jeff, and Adrienne Cachelin. 2018. "Critical Sustainability: Incorporating Critical Theories into Contested Sustainabilities." *Journal of Environmental Studies and Sciences* 8 (4): 518–525.

York, Richard, and Stefano B. Longo. 2017. "Animals in the World: A Materialist Approach to Sociological Animal Studies." *Journal of Sociology* 53 (1): 32–46.

Cultural Transmission

The culture of a group is the set of ideas, sometimes called rules or memes, that are held by its members (Dietz and Burns 1992). The culture of an individual guides their decisions and actions. For humans and many other species, culture is central to how they interact with other organisms and with the environment. Individuals acquire culture by observing and learning from others. This acquisition by observing and social learning is referred to as cultural transmission and is the core of learning for survival and sustainability (→ Learning for Survival).

Culture, like the genetics of a population, is shaped by evolution (→ Evolution), but the processes that shape the evolutionary dynamics of culture differ in important ways from those that shape the dynamics of genetic evolution (Richerson and Boyd 2005; Nichols et al. 2024). For example, in most sexually reproducing species half the genes come to offspring from each of two parents at conception. In contrast, cultural transmission comes from learning from many individuals and continues throughout the life course. Because of these dynamics, it is difficult for genetic evolution to produce individuals who are highly altruistic except under special conditions (→ Altruism). In contrast, it is much easier for cultural evolution to lead to altruism.

Recent research has revealed a great deal about the importance of cultural transmission and social learning in non-human animals (Brakes et al. 2025). The songs of many species of birds and whales are culturally transmitted and are crucial to social interactions. Chimpanzees and other primates have extensive culture, including knowledge traditions about how to construct and use tools in acquiring food (→ Animal Cultures). In long lived animals such as elephants, elders are repositories of culture, making their loss when targeted by hunting especially problematic (→ Learning for Survival). Conservation strategies and efforts to prevent extinction take into consideration the importance of cultural transmission in many species (→ Endangered Species Act). Early conservation efforts focused on preserving individuals and thus the genetic heritage of a species, including maintaining populations in zoos and sanctuaries (→ Zoos and Aquariums) (→ Sanctuaries). Current efforts emphasize preserving populations by protecting habitat. The survival of many species requires protecting their culture as well as their genetic diversity and habitats.

References

Brakes, Philippa, Lucy Aplin, Emma L. Carroll, Alison L. Greggor, Andrew Whiten, and Ellen C. Garland. 2025. "Animal Culture: Conservation in a Changing World." *Philosophical Transactions of the Royal Society B: Biological Sciences* 380 (1925): 20240127. https://doi.org/10.1098/rstb.2024.0127.

Dietz, Thomas, and Tom R. Burns. 1992. "Human Agency and the Evolutionary Dynamics of Culture." *Acta Sociologica* 35: 187–200.

Nichols, Ryan, Mathieu Charbonneau, Azita Chellappoo, et al. 2024. "Cultural Evolution: A Review of Theoretical Challenges." *Evolutionary Human Sciences* 6: e12. https://doi.org/10.1017/ehs.2024.2.

Richerson, Peter J., and Robert Boyd. 2005. *Not by Genes Alone: How Culture Transformed Human Evolution*. University of Chicago Press.

D

Defaunation

Defaunation is the loss of both local populations and species of wildlife as well as local declines in the abundance of individuals (Dirzo et al. 2014). While biodiversity loss is the reduction of living organisms (animals, plants, microbes) (→ Biodiversity), defaunation is a subset of biodiversity loss and refers specifically to the loss of animal life (fauna). There are four components to defaunation (Poulsen et al. 2023): 1) wildlife refers to all living animals who are not humans or domesticated; 2) defaunation incorporates the loss of populations and declines in abundance, so extinction and extirpation may result in some but not in all circumstances (→ Extinction); 3) defaunation can be driven by multiple causes such as habitat loss and degradation, wildfire, wildlife parasites or pathogens, human-wildlife conflict, roadkill, mesopredator release, hunting, the proliferation of invasive species, the capture of wildlife for the pet trade; and finally 4) the term defaunation is meant to be analogous to deforestation (→ Habitat Loss) (→ Zoonosis) (→ Conflict with Terrestrial Animals) (→ Conflict with Ocean Animals) (→ Road Ecology) (→ Mesopredator Release) (→ Empty Forest) (→ Invasive Animal Species) (→ Wildlife Trade).

The Living Planet Report of the World Wildlife Fund reports that global vertebrate populations declined by 69 percent between 1970 and 2018, with Latin America experiencing a 94 percent decline in the average abundance of vertebrates (da Silva Batista et al. 2025). Invertebrate populations are also declining. Monitoring data on a sample of 452 terrestrial invertebrate species documented a 45 percent decline in the abundance of individuals since 1970 (Dirzo et al. 2014; Young et al. 2016). Although emerging later than terrestrial defaunation by many thousands of years, animal loss in the global ocean has intensified with industrial fishing, the expansion of coastal human populations, bottom trawling, fish farming (→ Aquaculture), and increased human harvest of ocean animals (McCauley et al. 2015). For example, the mortality of sharks from fishing and the wasteful practice of shark finning "increased from at least 76 to 80 million sharks between 2012 and 2019," about 25 million of whom were threatened species (Worm et al. 2024, 225). In addition, fishing for oil and meat has exacerbated the defaunation of deepwater sharks and rays who have very low population growth rates because of their slow life

DOI: 10.4324/9781003528142-5

histories, long life spans, and low reproductive outputs (→ Learning for Survival) (Finucci et al. 2024).

Defaunation has also altered the diversity of microbes such as bacteria, viruses, and fungi, with strong consequences for ecosystem function. Interactions and feedback loops between top predators, through herbivores, to foundation species and their microbiomes must be considered to understand how defaunation affects all levels of biological organization as change radiates throughout ecosystems (Burkepile and Thurber 2019). For example, Pacific salmon transmit nutrients to aquatic and terrestrial systems during annual spawning; when they die, their carcasses deliver marine-derived nutrients to inland ecosystems altering both the diversity of benthic primary producers and riparian forest diversity (Burkepile and Thurber 2019).

Seddon et al. (2014, 406) note that "some progress in reversing defaunation is being achieved through the intentional movement of animals to restore populations ... from reinforcement and reintroduction to controversial conservation introductions that seek to restore populations outside their indigenous range or to introduce ecological replacements for extinct forms" (→ Rewilding). Most reinforcement projects focus on birds and mammals, and for translocations in general, few invertebrate, reptile, amphibian, or fish species are represented relative to their abundance in nature (Seddon et al. 2014). More than 485 animal species (mostly birds and mammals) have been reintroduced to their former distributions, and success has been achieved in reversing local extinctions of marine fauna (northern elephant seals, southern sea otters), freshwater fauna (Shasta River chinook salmon, Oregon chub), and terrestrial fauna (North American wolves, American crocodile, golden lion tamarin) (Young et al. 2016).

The consequences of systematic defaunation are twofold: 1) it alters basic ecosystem functions, including pollination, pest control, nutrient cycling and decomposition, water quality, and human health; and 2) it impacts evolutionary patterns because, for example, the loss of vertebrates often selects for body size, and smaller and short-lived organisms cannot fully replace the ecological services their larger counterparts provide (Dirzo et al. 2014). In addition to recognizing large mammal defaunation, understanding the ecosystem impact of functionally important fauna, such as nematodes, beetles, and bats, is critical – "if unchecked, Anthropocene defaunation will become ... a driver of fundamental global transformation in ecosystem functioning" (Dirzo et al. 2014, 406). Defaunation should be addressed as a major global environmental change with the same gravity as deforestation, pollution, and climate change (→ Global Environmental Change) (→ Pollution) (→ Climate Change) (Young et al. 2016, 333).

References

Burkepile, Deron E., and Rebecca Vega Thurber. 2019. "The Long Arm of Species Loss: How Will Defaunation Disrupt Ecosystems Down to the Microbial Scale?" *Bioscience* 69 (6): 443–454.

da Silva Batista, Gabriela, Carlos R. Brocardo, Arlison B. Castro, et al. 2025 "Defaunation Disrupts the Behavior of Large Terrestrial Vertebrates, Impacting Ecological Functions in the Amazon." *Global Ecology and Conservation* 59: e03522. https://doi.org/10.1016/j.gecco.2025.e03522
Dirzo, Rodolfo, Hillary S. Young, Mauro Galetti, et al. 2014. "Defaunation in the Anthropocene." *Science* 345 (6195): 401–406.
Finucci, Brittany, Nathan Pacoureau, Cassandra L. Rigby, et al. 2024. "Fishing for Oil and Meat Drives Irreversible Defaunation of Deepwater Sharks and Rays." *Science* 383: 1135–1141.
McCauley, Douglas J., Malin L. Pinsky, Stephen R. Palumbi, James A. Estes, Francis H. Joyce, and Robert R. Warner. 2015. "Marine Defaunation: Animal Loss in the Global Ocean." *Science* 347 (6219): 1255641–1.
Poulsen, John R., Vincent Maicher, Halina Malinowski, and Camille DeSisto. 2023. "Situating Defaunation in an Operational Framework to Advance Biodiversity Conservation." *BioScience* 73: 721–727. https://doi.org/10.1093/biosci/biad079.
Seddon, Philip J., Critistine J. Griffiths, Pritpal S. Soore, and Doug P. Armstrong. 2014. "Reversing Defaunation: Restoring Species in a Changing World." *Science* 345 (6195): 406–412.
Worm, Boris, Sara Orofino, Echelle S. Burns, et al. 2024. "Global Shark Fishing Mortality Still Rising Despite Widespread Regulatory Change." *Science* 383: 225–230.
Young, Hillary S., Douglas J. McCauley, Mauro Galetti, and Rodolfo Dirzo. 2016. "Patterns, Causes, and Consequences of Anthropocene Defaunation." *Annual Review of Ecology, Evolution, and Systematics* 47: 333–358.

Disability

Disability, in human and non-human animals, is defined as "an array of cognitive, sensory, and physical differences and capabilities that may substantially alter one or more major life functions for human and nonhuman animals either permanently or temporarily" (Whitley et al. 2025a, 1). This definition is based on the social disability model in disability studies, which asserts that disability arises not just from individual limitations, but from the limitations placed on individuals through discriminatory attitudes, policies, and practices (Shakespeare 2013; Oliver and Barnes 2012). Disability studies challenges dominant constructions of "normal" ability by illustrating how cultural ideas of productivity, independence, and merit systematically marginalize individuals who move, think, or communicate differently.

Globally, thousands of animals with disabilities are living in homes, zoos, aquariums, sanctuaries, and in the wild (→ Domestic Animals) (→ Zoos and Aquariums) (→ Sanctuaries) (→ Wilderness Animals). Environmental and sustainability scholars have begun to recognize that disability should not be peripheral to discussions of ecological sustainability, and by keeping disability as a peripheral topic ableism and speciesism are reinforced (→ Speciesism). Ableism is a set of beliefs and practices embedded in culture that normalize the devaluation and discrimination of those with disabilities (Center for Disability Rights 2025). Like disability, ableism is contextual. Survey research shows that people often disagree about what disability means and what should be considered a disability. The cognitive abilities and sentience of non-human animals are often debated, and this can pose additional challenges for discussing psychiatric and cognitive disabilities in non-human animals (→ Sentience) (→ Cognition) (Bekoff 2015).

Disability and animal studies scholar Sunaura Taylor (2017) contends that cultural norms privileging autonomy, strength, and productivity inform the ways society devalues both human and non-human animals with disabilities, often constructing them as weak, dependent, and without purpose, leading many institutions to choose euthanasia over long-term care and management. Yet, "we have no idea how other animals comprehend physical or cognitive difference" (Taylor 2017, 24). Animals with disabilities are not only important members of communities and ecosystems, but also their visibility can help to inform positive attitudes around disability and sustainability broadly (→ Ecosystems) (Whitley et al. 2025b). Furthermore, queer ecology has contributed to disability-informed scholarship on animals by challenging fixed assumptions about bodily norms and ecological suitability (→ Queer Ecology), thereby reframing animals with disabilities as routine rather than exceptional (Whitley et al. 2025a).

Basil, a Virginia opossum, was one of the most famous public animals with a disability until his passing in 2025 (Godfrey 2023). Two years before his death, Basil was attacked by a predator and ended up at a local rehabilitation center. Because Basil's injuries were too extensive for him to survive in the wild, he became a resident at the Smithsonian's National Zoo & Conservation Biology Institute (also known as the National Zoo) in Washington, DC. The National Zoo thought that Basil's story and disability could raise awareness about the importance of opossums in ecosystems and potentially help with conservation and sustainability efforts, so their marketing focused on depicting Basil as an individual with a unique personality who happened to have a disability. His personality and abilities were centered over the tragedy, trauma, and disability. This approach proved effective, and Basil's story was subsequently circulated through both local and national media outlets. Research shows that simple depictions of animals with disabilities that center individual personalities over trauma and disability are most effective at enhancing positive perceptions of animals and people with disabilities (Whitley et al. 2025b).

Animal welfare and conservation organizations have long grappled with how to discuss animals with disabilities in ways that honor the animals themselves, advance sustainability and conservation goals, and avoid reinforcing negative perceptions of either disability or the caretaking organization. The Americans with Disabilities Act National Network (ADA 2024) provides guidelines for talking about people with disabilities, but until recently there were no such guidelines for talking about animals with disabilities. Whitley and colleagues (2025a) crafted guidelines to better help animal organizations talk about animals with disabilities to enhance positive perceptions of both non-human animals and humans with disabilities. The guidelines and evolving research are intended to not only inform discourse about animals with disabilities, but also to confront broader devaluation of those with disabilities. In doing so, they highlight the connection between disability, sustainability, and the shared reality that all ecosystems, including human and non-human communities, include members with disabilities.

References

ADA. 2024. "Guidelines for Writing About People with Disabilities." https://adata.org/factsheet/ADANN-writing.

Bekoff, Marc. 2015. "Psychological Disorders in Animals: A Review of What We Know." *Psychology Today*, September 9. www.psychologytoday.com/us/blog/animalemotions/201509/psychological-disorders-in-animals-review-what-we-know.

Center for Disability Rights. 2025. "#Ableism." Center for Disability Rights. https://cdrnys.org/blog/uncategorized/ableism/.

Godfrey, Elaine. 2023. "Basil the Opossum Has One Eye, a Big Heart, and a Job to Do." *The Atlantic*, September 28. www.theatlantic.com/science/archive/2023/09/one-eyedopossum-national-zoo-backyard-animals/675482/.

Oliver, Michael, and Colin Barnes. 2012. *The New Politics of Disablement*. Bloomsbury.

Shakespeare, Tom. 2013. *Disability Rights and Wrongs Revisited*. Routledge.

Taylor, Sunaura. 2017. *Beasts of Burden: Animal and Disability Liberation*. The New Press.

Whitley, Cameron T., Marta Burnet, Em Sherwood, et al. 2025a. "Increasing Positive Perception of Disability through Depictions of Animals with Disabilities." *Animals* 15 (13): 1861.

Whitley, Cameron T., Denny Dulaney, Brooklyn Wehr, Marta Burnet, Alexander Jones, and Courtney Cordova. 2025b. "Guidance for Depicting Animals with Disabilities." *Curator: The Museum Journal* 69 (1): 117–121.

Disasters

Disasters are any sudden events or accidents where a hazard (natural or human-made) interacts with social conditions to create widespread destruction and devastation that exceeds a human or non-human animal community's capacity to manage. Hazards, in a disaster context, are natural or human-made processes or phenomenon that pose a risk to sentient beings (human and non-human animals), flora, and/or non-sentient objects such as property (→ Sentience) (→ Property). Hazards only become disasters when they impact human or non-human animal communities. Disasters are often conceptualized as naturally occurring phenomena, such as earthquakes, hurricanes, volcanic eruptions, or floods, that may be exacerbated by human activities like urban development. Yet, human-made hazards, including oil spills and nuclear accidents, can also precipitate disasters with far-reaching social and ecological consequences. Coverage of animal loss, and disruptions to human-animal relationships, remain nearly absent in media coverage of disasters. In most cases, non-sentient objects and sentient animal loss are quantified economically, while the loss of human life is categorized numerically or emotionally. Increasingly, scholars emphasize that disasters not only impact humans but also entire ecosystems, including the animals who live alongside humans, depend on them, or inhabit shared environments (→ Ecosystems).

Some disasters, such as the 2011 Fukushima Nuclear Disaster, occur because a naturally occurring event leads to a technological failure. On March 11, 2011, an earthquake and resulting tsunami led to the failure of the electrical grid and all backup energy at the Fukushima Daiichi Nuclear Power Plant in Okuma, Fukushima, Japan, which caused a major nuclear accident. The inability to cool reactors after the shutdown led to the release of radioactive contaminants into the surrounding environment. Animals were immediately exposed to the dangers of the earthquake, tsunami, and flooding. This led to the immediate loss of animal life, mass starvation, euthanasia of farmed animals, and trauma to companion animals (Mattes 2017; Mattes and Whitley 2021). Enduring impacts remain, particularly those related to radiation and genetic degradation, which can hinder the reproductive success of some wild species (Møller et al. 2012; Møller et al. 2013).

Discussions of animals in disasters have predominantly focused on companion animals, as these animals share the most intimate and interdependent relationships with humans (→ Domestic Animals). Concerns about the welfare of human-animal relationships in disasters even prompted bipartisan legislation designed to better protect pets and their owners in extreme events. The Pets Evacuation and Transportation Standards (PETS) Act was enacted in 2006 due to the widespread suffering of companion animals and humans during and after Hurricane Katrina made landfall in 2005 (→ Pets Evacuation and Transportation Standards Act). The PETS Act requires emergency plans to account for companion animals. Even with the PETS Act, companion animals are disproportionately impacted by disasters (Irvine 2009). For instance, as Irvine and Andre (2023) show, in the Marshall Urban Wildfire in Colorado in 2021, two humans perished, while over 1000 companion animals died. Scholars argue that better notification systems and the allocation of support resources are needed to help evacuate companion animals from homes that are under threat (Irvine and Andre 2023).

The effects of disasters extend far beyond companion animals, encompassing farmed animals (also known as livestock or agricultural animals) as well (→ Livestock Production). In 2019, nearly one million farmed animals were reported to have died due to disasters in the US alone (Verma and Prem 2022). Prior to that, in 2018 flooding from Hurricane Florence led to the drowning of millions of chickens and pigs in industrial barns (Stoddard and Hovorka 2019). The exact number of animals who perished in Hurricane Florence is unknown. The vulnerability of farmed animals is largely due to these animals having few protections and to legislation that pays contract producers for animal losses through programs like the Livestock Indemnity Program (LIP). While most disaster assistance programs require the demonstration of a risk-management plan or private insurance, the LIP does not (Verma and Prem 2022). In 2019, $58.5 million was paid out for farmed animal loss due to disasters (Verma and Prem 2022). These payments can incentivize animal loss over animal rescue which can be challenging and costly for contract producers.

When disasters occur, the impacts on wilderness animals are often devastating, but largely overlooked (→ Wilderness Animals). Many species face immediate threats such as injury, displacement, and habitat loss. Others struggle with long-term effects like changes to migration routes and food availability. Events like wildfires, hurricanes, and oil spills can decimate entire populations and forever alter ecosystems. However, reduced human activity after disasters allows certain species to recover and expand territory.

Disasters reveal the interconnectedness of ecological, social, and economic systems, highlighting that sustainability cannot be achieved without addressing multispecies impacts (→ Multispecies Futures). When hazards devastate both human and non-human communities, they expose the fragility of ecosystems and the perils of unsustainable development and land use (→ Habitat Loss). Understanding how disasters impact animals and human-animal relationships offers insight for creating resilient, sustainable recovery strategies that help to restore and protect human and non-human animal ecosystem communities.

References

Irvine, Leslie. 2009. *Filling the Ark: Animal Welfare in Disasters*. Temple University Press.

Irvine, Leslie, and Casara Andre. 2023. "Pet Loss in an Urban Firestorm: Grief and Hope after Colorado's Marshall Fire." *Animals* 13 (3): 416.

Mattes, Seven. 2017. "The Shared Vulnerability and Resiliency of the Fukushima Animals and Their Rescuers." In *Responses to Disasters and Climate Change: Understanding Vulnerability and Fostering Resilience*, edited by Michelle Companion and Miriam Chaiken, 103–116. Taylor & Francis.

Mattes, Seven, and Cameron T. Whitley. 2021. "Entangled Impacts: Human-Animal Relationships and Energy Development." In *Energy Impacts: A Multidisicplinary Exploration of North American Energy Development*, edited by Jeffrey B. Jacquet, Julia H. Haggerty, and Gene L. Theodori, 54–73. University Press of Colorado.

Møller, Anders Pape, Atsushi Hagiwara, Shin Matsui, et al. 2012. "Abundance of Birds in Fukushima as Judged from Chernobyl." *Environmental Pollution* 164: 36–39.

Møller, Anders Pape, Satoe Kasahara, Shin Matsui, et al. 2013. "Differences in Effects of Radiation on Abundance of Animals in Fukushima and Chernobyl." *Ecological Indicators* 24: 75–81.

Stoddard, Elisabeth A., and Alice Hovorka. 2019. "Animals, Vulnerability and Global Environmental Change: The Case of Farmed Pigs in Concentrated Animal Feeding Operations in North Carolina." *Geoforum* 100: 153–65.

Verma, Subhash, and Hansen Thambi Prem, eds. 2022. *Management of Animals in Disasters*. Springer.

Domestic Animals

Domestic animals are non-human animal species who coevolved with humans over a long period of time. They develop behavioral, physiological, and morphological traits that enhance their compatibility with human needs, often resulting in reduced fear, dietary adaptability, and social tolerance. Domesticated animals include a wide range of species from dogs and cats to bees (Larson and Fuller 2014).

Most animal domestication occurred roughly 9–10,000 years ago (Adhikari 2023). Dogs are the exception. While the where, when, and even the frequency of dog domestication has long been the subject of debate, Evin et al. (2025) have made a major contribution to the clarification of when the domestic dog emerged with a morphology different from wolves. In a large-scale morphometric study examining 643 canid crania spanning approximately 50,000 years, Evin and colleagues report that cranial size and shape consistent with domestic dogs and different from wolves emerged around 11,000 years ago. And a recent advanced genomic analysis of the DNA of 216 canid remains by Bergström and colleagues (2026) has now pushed the genetic history of dogs in Europe back to 14,200 BP.

While all companion animals are domestic, not all domestic animals are companions. Contemporary scholarship emphasizes that domestication is not solely a biological process, but also a social and cultural one that is shaped by human values, economic systems, and environmental conditions (Zeder 2016). Budiansky (1992) and others have long argued that domestication emerged through a process of self-selection in which opportunistic animals were drawn to human-provided food sources. Over successive generations, these populations became increasingly tractable, in part through the retention of juvenile behavioral and developmental traits in contexts where predation pressures were reduced (Clutton-Brock 2012).

Domestic animals occupy a unique position at the intersection of human society, ecological systems, and global food and companionship networks. These dynamics can be illustrated through the example of cattle. Cattle were once fully wild animals, members of the now extinct aurochs (*Bos primigenius*), who ranged widely across Europe, Asia, and North Africa before their domestication roughly 10,000 years ago. Over time, selective breeding through domestication created a species who diverged substantially from their wild ancestors in morphology, behavior, and ecological flexibility (Campbell 2025). Because the aurochs was driven to extinction in the seventeenth century, no wild population exists today. Instead, domestic cattle exist entirely in human-managed systems for production (→ Livestock Production). This example underscores how domestication not only alters animal bodies and behaviors but can erase the wild origins of species. Although research demonstrates that cattle are sentient, having complex social lives and requiring environments that permit movement, social interaction, and nutritionally appropriate foraging, such conditions are largely absent in most livestock production systems (→ Sentience) (→ Cognition) (Miller-Cushon 2024). However, there have been advances in cattle welfare over the past 50 years (Grandin 2025), and some argue for a livestock production system that enhances intersubjective relations between humans and animals. For example, Jocelyn Porcher's (2017) research on dairy cows shows that animals actively participate in the work they perform, communicating with others, forming bonds, negotiating rules, taking initiative, and seeking recognition for their contributions.

Domestic animals are deeply tied to sustainability because they are central to global food production, land use, and climate dynamics. Livestock production contributes significantly to greenhouse gas emissions, freshwater consumption, and habitat conversion, making domestic animals a focal point in climate, biodiversity, and sustainability debates (Havlík et al. 2014). At the same time, sustainable models, such as rotational grazing, reduced stocking densities, plant-based transitions, and integrated crop-animal systems, highlight pathways for reducing environmental harm (→ Veganism) (→ Vegetarianism). As societies confront climate change and ecological decline, the role of domestic animals invites critical reflection on how human-animal relationships can evolve to support more resilient, equitable, and ecologically responsible futures (→ Multispecies Futures).

References

Adhikari, Ram Chandra. 2023. "An Overview of the Past of Animal Domestication." *Medha: A Multidisciplinary Journal* 6 (1): 55–65.

Bergström, Anders, Anja Furtwängler, Sarah Johnston, et al. 2026. "Genomic History of Early Dogs in Europe." *Nature* 651: 986–994. https://doi.org/10.1038/s41586-026-10112-7.

Budiansky, Stephen. 1992. *The Covenant of the Wild: Why Animals Chose Domestication: With a New Preface*. Yale University Press.

Campbell, Michael O'Neal. 2025. *Cattle, Their Predators and Geomatics Research*. Springer Nature.

Clutton-Brock, Juliet. 2012. *Animals as Domesticates: A World View through History*. Michigan State University Press.

Evin, Allowen, Carly Ameen, Colline Brassard, et al. 2025. "The Emergence and Diversification of Dog Morphology." *Science* 390 (6774): 741–744.

Grandin, Temple. 2025. "The Evolution of Cattle Welfare over the Past Few Decades." *Veterinary Clinics: Food Animal Practice* 41 (3): 471–479.

Havlík, Petr, Hugo Valin, Mario Herrero, et al. 2014. "Climate Change Mitigation through Livestock System Transitions." *Proceedings of the National Academy of Sciences* 111 (10): 3709–3714.

Larson, Greger, and Dorian Q. Fuller. 2014. "The Evolution of Animal Domestication." *Annual Review of Ecology, Evolution, and Systematics* 45 (1): 115–136.

Miller-Cushon, Emily K. 2024. "Current Research Considering Social Behavior to Improve the Welfare of Commercially Raised Dairy Calves." *JDS Communications* 5 (3): 264–269.

Porcher, Jocelyn. 2017. "Animal Work." In *The Oxford Handbook of Animal Studies*, edited by Linda Kalof, 302–318. Oxford University Press.

Zeder, Melinda A. 2016. "Domestication as a Model System for Niche Construction Theory." *Evolutionary Ecology* 30 (2): 325–348.

E

Ecocentrism

Ecocentrism "argues for the intrinsic value of all elements in nature, suggesting that ecosystems and non-human entities deserve moral consideration independent of their utility to humans" (Kalita and Barsha 2024, 448). This perspective contrasts with anthropocentrism that argues human interests take priority over the natural world and the environment has value only insofar as it benefits humans (→ Anthropocentrism). A shift from anthropocentric to ecocentric thinking is necessary for sustainable living in the face of unprecedented environmental crises and biodiversity loss (→ Global Environmental Change) (→ Biodiversity) (Kalita and Barsha 2024).

Ecocentrism's roots lie in multiple perspectives that embrace the intrinsic value of Earth, such as Indigenous cultures, Aldo Leopold's 1949 "Land Ethic," Arne Naess' 1973 "Deep Ecology," Stan Rowe's 1994 arguments for ecocentrism as a resolution of the environmental crisis (Washington et al. 2017), and by extension feminist care theory (→ Ecofeminism) (→ Indigenous and Decolonial Perspectives) (Donovan 2022; Fitzgerald 2019). In their "Manifesto for Earth," Mosquin and Rowe (2004) provide six core principles of ecocentrism: 1) The ecosphere is the center of value for humanity, 2) the creativity and productivity of Earth's ecosystems depend on their integrity, 3) the Earth-centered worldview is supported by natural history, 4) ecocentric ethics are grounded in awareness of our place in nature, 5) an ecocentric worldview values diversity of ecosystems and cultures, and 6) ecocentric ethics support social justice.

Gray et al. (2020) argue that making conservation decisions based on anthropomorphism leads to irreversible harm to non-humans. Thus, a major aspect of moving ecocentrism into practice must be adequate political representation for non-humans and ecosystems. Only such representation sufficiently addresses their intrinsic value as entities with interests (→ Ecodemocracy).

Research on human values shows that many individuals endorse ecocentric values (→ Values). Ecocentric values are also central to the successful development of international instruments for conservation and the implementation of governmental conservation policies (Taylor et al. 2020). For example,

DOI: 10.4324/9781003528142-6

ecocentrism has inspired visionary proposals to maintain biodiversity, such as the Global Biodiversity Framework that commits 195 nations and the European Union to conserve 30 percent of land and sea areas by 2030 and the promotion of the goal of protecting half of the planet by 2050 (→ Sustainable Development Goals) (Shiffman 2020; Taylor et al. 2020). Recent research shows that strong majorities of the population in eight countries support these goals (Argentina, Brazil, India, Indonesia, South Africa, Spain, Sweden, and the United States) (Michaelsen et al. 2025).

References

Donovan, Josephine. 2022. *Animals, Mind, and Matter: The Inside Story*. Michigan State University Press.

Fitzgerald, Amy J. 2019. *Animal Advocacy and Environmentalism*. Polity Press.

Gray, Joe, Anna Wienhues, Helen Kopnina, and Jennifer DeMoss. 2020. "Ecodemocracy: Operationalizing Ecocentrism through Political Representation for Non-humans." *The Ecological Citizen* 3: 166–177.

Kalita, Namita, and Jigyasa Barsha. 2024. "Ecocentrism vs. Anthropocentrism: Reimagining Humanity's Relationship with Nature." *International Journal of Humanities and Social Science and Management* 4 (6): 448–450.

Michaelsen, Patrik, Aksel Sundström, and Sverker C. Jagers. 2025. "Mass Support for Conserving 30% of the Earth by 2030: Experimental Evidence from Five Continents." *Proceedings of the National Academy of Sciences* 122 (35): e2503355122.

Mosquin, Ted, and Stan Rowe. 2004. "A Manifesto for Earth." *Biodiversity* 5 (1): 3–9.

Shiffman, David. 2020. "An Ambitious Strategy to Preserve Biodiversity." *Scientific American*, December 1. www.scientificamerican.com/article/an-ambitious-strategy-to-preserve-biodiversity/.

Taylor, Bron, Guillaume Chapron, Helen Kopnina, Ewa Orlikowska, Joe Gray, and John J. Piccolo. 2020. "The Need for Ecocentrism in Biodiversity Conservation." *Conservation Biology* 34 (5): 1089–1096.

Washington, Haydn, Bron Taylor, Helen Kopnina, Paul Cryer, and John J. Piccolo. 2017. "Why Ecocentrism Is the Key Pathway to Sustainability." *The Ecological Citizen* 1 (1): 35–41.

Ecocriticism

Ecocriticism is an interdisciplinary field that examines the representations of nature, environment, and ecological relationships in literature, film, and other cultural texts (Glotfelty and Fromm 1996). Emerging from literary studies in the 1990s, ecocriticism explores how cultural narratives shape human understandings of the natural world and the ethical responsibilities that follow from those interpretations. It not only considers how environments are depicted, but also how power, ideology, and historical context influence those depictions. Contemporary ecocriticism often includes perspectives from environmental studies, environmental humanities, postcolonial studies, and ecofeminism, inviting readers to question anthropocentric assumptions and to recognize the interconnectedness of human and non-human life (→ Anthropocentrism) (→ Biocentrism) (Shamim et al. 2025).

A useful example of ecocriticism's relevance to non-human animals can be seen in analyses of animal representations in literature. Scholars have examined how novels, memoirs, and nature writing portray animals as symbolic figures, moral subjects, or agents with their own perspectives. For instance, studies of whale narratives, from Herman Melville's *Moby-Dick* to contemporary works like Philip Hoare's *The Whale: In Search of Giants of the Sea*, investigate how cultural stories about whales shape public attitudes and values toward conservation and whaling (→ Values) (Borcan 2025). Ecocritical readings of such texts illuminate the ways that literary depictions can either reinforce exploitative relationships or encourage empathy and ethical engagement with non-human animals (→ Empathy).

An important strain of ecocriticism has emerged around comic books and graphic novels. While once considered "trash" reading, comic books and graphic novels are now considered an important part of popular culture. During the 1950s "funny animals" who were strongly anthropomorphized became very popular, even as superheroes, horror, and crime comics were being cancelled (→ Critical Anthropomorphism). Most of these characters, such as the Fox and Crow and Bugs Bunny, were depicted entirely for humor. Others, most notably Carl Barks' Uncle Scrooge and Donald Duck, while often funny, also involved adventures. At its peak, *Walt Disney's Comics and Stories*, featuring Barks' Donald Duck as well as Mickey Mouse, had a monthly circulation of 3 million and reached 6 percent of US households (Immerwahr 2020).

Barks' stories are noted for interrogating assumptions about our interactions with other species and with the environment – precursors to sustainability concerns (Andrae 2006; Immerwahr 2020). In one of the most famous stories, Scrooge McDuck tries to buy a vast wilderness area. He is confronted by both the Indigenous humans and by squirrels, birds, fish, and other species. In a famous speech, Chief Brave Eagle, surrounded by various woodland species, says "None could sign away these woodlands, None could have the right or reason, But the chiefs of all the brothers in a powwow with the seasons" (Barks 1957). It is a clear and early statement of ecodemocracy (→ Ecodemocracy). The primary readers of Barks' work at the time were the baby boomer cohort that would later shape the environmental movement.

More recent comics have taken up animal and sustainability issues. Notably, during Grant Morrison's authorship, the superhero Animal Man became vegan and engaged in support of animal liberation actions, such as protecting foxes from hunters and freeing laboratory animals (→ Veganism) (→ Abolitionism). Animal Man's power is to take on the abilities of non-human animals such as speed, strength, and flying, engaging him with non-humans and raising his empathy for them. It spurred extensive scholarship in critical animal studies that not only examines comic books but also graphic novels and the long history of animal characters in popular newspaper comic strips (→ Critical Animal Studies) (Mahmutovic et al. 2021; Schatz and Parson 2017; Alaniz 2025).

Ecocriticism contributes to sustainability debates by highlighting how cultural stories influence environmental behavior, policy, and moral consideration. Because sustainability requires shifts not only in technology and resource use, but also in values, ecocriticism helps identify the narrative frameworks that either support or hinder more just and ecologically attuned futures (Radhakrishnan 2025). When ecocritics analyze how literature represents animals, whether as commodities, companions, or sentient beings, they uncover underlying assumptions that shape how societies treat other species (→ Sentience) (Yadav and Sinha 2024). By challenging anthropocentrism and encouraging more inclusive ecological imaginaries, ecocriticism supports efforts to rethink multispecies coexistence and to build sustainability approaches that consider the well-being of both humans and nonhuman animals (→ Multispecies Futures). Ecocritical scholarship is poised to play an increasingly significant role in shaping public understanding of environmental issues and fostering more sustainable ways of living in the Anthropocene (→ Anthropocene) (Shamim et al. 2025).

References

Alaniz, José. 2025. *Comics of the Anthropocene: Graphic Narrative at the End of Nature*. University Press of Mississippi.

Andrae, Tom. 2006. *Carl Barks and the Disney Comic Book: Unmasking the Myth of Modernity*. University Press of Mississippi.

Barks, Carl. 1957. "Land of the Pygmy Indians." *Uncle Scrooge*, June.

Borcan, Andrei Dimitrie. 2025. "Ecocritical Tropes in Herman Melville's *Moby Dick or The Whale*." *Journal of Romanian Literary Studies* (40): 393–401.

Glotfelty, Cheryll, and Harold Fromm. 1996. *The Ecocriticism Reader: Landmarks in Literary Ecology*. University of Georgia Press.

Immerwahr, Daniel. 2020. "Ten-Cent Ideology: Donald Duck Comic Books and the U.S. Challenge to Modernization." *Modern American History* 3 (1): 1–26. https://doi.org/10.1017/mah.2020.4.

Mahmutovic, Adnan, David Coughlan, and Stephen Blake Ervin. 2021. "*Ecce Animot*: Or, The Animal Man That Therefore I Am." *ImageText* 8 (2). https://imagetextjournal.com/ecce-animot-or-the-animal-man-that-therefore-i-am/.

Radhakrishnan, Ganesh V. 2025. "The Role of Literature in Shaping Environmental Consciousness: Ecocritical Perspectives in Modern English Literature." *Advances in Consumer Research* 2 (1): 24.

Schatz, J.L., and Sean Parson, eds. 2017. *Superheroes and Critical Animal Studies*. Bloomsbury Publishing.

Shamim, Mohd, Usha Sahu, and Vivek Kumar Dwivedi. 2025. "Ecocriticism in Literature: Examining Nature in Contemporary Fictions." *International Journal of Social Impact* 10 (2): 405–415.

Yadav, Chesta, and Jyotsna Sinha. 2024. "Exploring the Evolution of Ecocriticism: A Bibliometric Study and Literature Review." *Multidisciplinary Reviews* 7 (12): 2024304.

Ecodemocracy

Ecodemocracy is an ecocentrically modified form of democracy where communities use "decision-making systems that respect the principles of human democracy while explicitly extending valuation to include the intrinsic value of non-human nature, with the ultimate goal of evaluating human wants equally to those of other species and the living systems that make up the ecosphere" (Gray and Curry 2016, 21). Ecodemocracy holds that consideration should be given to the intrinsic value of both the biotic parts of the biosphere as well as the abiotic parts, such as landforms, rivers, and soil (→ Ecocentrism) (Gray and Curry 2020). Ecodemocracy can be thought of as an extension of arguments for deliberative democracy and deliberative ethics that emphasize decision-making processes where all interested and impacted parties have a voice, and the process itself is fair and respectful (Dietz 2023).

The challenge to ecodemocracy is how to ensure the representation of non-human nature, in other words, "who speaks for wolf?" Some scholars argue that the most logical way to ensure that the intrinsic value of non-human nature is considered in socio-political systems is a deliberative process that assigns stakeholder status and voting rights to non-humans that would be carried out by human proxies (Gray and Curry 2016). Others argue that animals have the right not just to indirect representation by human trustees, but to "direct political voice and participation through processes of self-representation to the extent this is possible" (Donaldson 2020, 710). Justice for the rights of animals should be situated in a political theory of democracy, citizenship, and sovereignty that gives animals political status and inclusion, such as citizenship for domestic animals, sovereignty for wild animals, and denizenship for liminal animals (→ Rights) (→ Domestic Animals) (→ Wilderness Animals) (→ Liminal Animals) (Donaldson and Kymlicka 2011). Political participation for domestic animals could be achieved through the creation of a multispecies public commons (→ Multispecies Futures). The commons would be governed by all who inhabit them, an "animal agora" that would release animals from confinement, enclosure, and segregation into public spaces where they are free to act outside the control of humans (→ Agency) (Donaldson 2020). For example, borrowing from critical disability theory, the mere presence of domestic animals in public spaces (such as dogs on trains or in restaurants) can change people's ideas about who belongs and who matters, disrupting long-standing patterns of oppression (→ Disability) (Donaldson 2020, 712). Finally, Meijer (2019) argues that it is unfortunate that humans must speak for animals because it perpetuates the view that other animals are mute and incapable of communicating. In fact, animals "have their own perspective on life and their own ways of communicating this to others of their own and other species" (→ Animal Languages), and their political participation involves finding ways for humans to communicate with other animals and learn their attitudes to the questions that concern their lives (→ Critical Anthropomorphism) (Meijer 2019, 5–6).

The concept of giving personhood with certain rights and moral consideration to non-human entities emerged in US politics in the early seventies (→ Rights). In the 1972 case of *Sierra Club vs. Morton* (405 U.S. 727), Supreme Court Justice William O. Douglas argued that "environmental objects" should have legal personhood, leading to discussions of "Should Trees Have Standing?" but that idea is not part of established US law (Stone 1985). Some countries, however, have granted legal rights to non-human nature. In 2011, Bolivia granted all nature rights equal to that of humans, and in 2017, New Zealand granted legal personhood to the Whanganui River and its surrounding ecosystem. And in a sweeping motion to grant the right to "exist, flourish, and evolve," Ecuador has ruled that wild animals possess distinct legal rights and that mining in a protected cloud forest violates the rights of nature (although hunting, gathering, and forestry are allowed if performed within pre-existing laws and are carried out in ways that limit suffering) (Koumoundouros 2022). It is no coincidence that these cases of ecodemocratic legislation have been adopted by nations with Indigenous cultures that have greater attachment to the natural world than exists in more Westernized countries (→ Indigenous and Decolonial Perspectives) (Gray and Curry 2016). Granting rights to non-human animals and the natural world fosters an ecocentric perspective that is critical for sustainability (→ Ecocentrism).

References

Dietz, Thomas. 2023. *Decisions for Sustainability: Facts and Values*. Cambridge University Press.

Donaldson, Sue. 2020. "Animal Agora: Animal Citizens and the Democratic Challenge." *Social Theory and Practice* 46 (4): 709–735.

Donaldson, Sue, and Will Kymlicka. 2011. *Zoopolis: A Political Theory of Animal Rights*. Oxford University Press.

Gray, Joe, and Patrick Curry. 2016. "Ecodemocracy: Helping Wildlife's Right to Survive." *Ecos* 37 (1): 18–27.

Gray, Joe, and Patrick Curry. 2020. "Ecodemocracy and Political Representation for Non-human Nature." In *Conservation: Integrating Social and Ecological Justice*, edited by Helen Kopnina and Haydn Washington, 155–166. Springer.

Koumoundouros, Tessa. 2022. "The First Country in the World has Given Legal Rights to Individual Wild Animals." *Science Alert: Nature*. www.sciencealert.com/the-first-country-in-the-world-has-given-legal-rights-to-individual-wild-animals.

Meijer, Eva. 2019. *When Animals Speak: Toward an Interspecies Democracy*. New York University Press.

Stone, Christopher D. 1985. "Should Trees Have Standing Revisited: How Far Will Law and Morals Reach—A Pluralist Perspective." *Southern California Law Review* 59 (1): 1–154.

Ecofeminism

Ecofeminism is an interdisciplinary framework that links the domination of women and other marginalized communities with the domination of nature, arguing that these forms of oppression share historical, cultural, and political roots (Öztürk 2020). Emerging in the late twentieth century, ecofeminism critiques dualistic thinking, such as human/animal, culture/nature, and masculine/feminine, that has justified social hierarchies and environmental exploitation (Gaard 2015). While diverse in its strands, ecofeminism generally maintains that gender justice and ecological well-being are interconnected, and that addressing environmental crises requires challenging systems such as patriarchy, colonialism, and capitalism that devalue both women's labor and the natural world (→ Colonialism). Work on environmental concern and risk has documented that white men tend to be less concerned with the environment than others, and one of the leading social psychological theories of environmentalism suggests that this is in part because of gender differences in altruism (→ Altruism) (Stern et al. 1993).

The relevance of ecofeminism to non-human animal concerns is found in broad critiques of exploitation across all industries, but specifically for industrial livestock production (→ Livestock Production). Ecofeminist scholars argue that the same logic that positions women as subordinate also permits the treatment of animals as disposable commodities. Carol J. Adams' (1990/2024) influential work on the "absent referent" demonstrates how language and cultural practices obscure the individuality of animals, particularly the promotion of carnism and meat production, making violence against them seem normal or invisible (→ Carnism). Similarly, ecofeminist analyses of dairy production highlight how female animals' reproductive capacities are exploited in systems designed for profit rather than care. These examples reveal how ecofeminism draws attention to the gendered dimensions of animal use and encourages more empathetic, relational approaches to interspecies ethics (→ Empathy).

Ecofeminism contributes to sustainability by emphasizing that ecological health depends on dismantling social hierarchies that harm both humans and non-human species. Recent scholarship argues that sustainable futures require valuing care, reciprocity, and multispecies justice rather than extractive relationships (→ Multispecies Futures) (Whyte et al. 2018). By foregrounding the voices of women, queer and Indigenous scholars, and communities disproportionately affected by environmental degradation, ecofeminism broadens what counts as environmental knowledge and ethical responsibility. When applied to non-human animals, ecofeminism encourages policies and practices that recognize animals' agency, promote coexistence, and reduce harm, ranging from promoting vegetarianism and veganism to habitat protection. In this way, ecofeminism offers a holistic approach to sustainability that intertwines social equity, environmental justice, and the well-being of all species (→ Environmental Justice) (→ Biocentrism).

References

Adams, Carol J. 1990/2024. *The Sexual Politics of Meat: 35th Anniversary Edition: A Feminist Vegan Critical Theory*. Bloomsbury.

Gaard, Greta. 2015. "Ecofeminism and Climate Change." *Women's Studies International Forum* 49 (March-April): 20–33.
Öztürk, Yıldız Merve. 2020. "An Overview of Ecofeminism: Women, Nature and Hierarchies." *The Journal of Academic Social Science Studies* 13 (81): 705–714.
Stern, Paul C., Thomas Dietz, and Linda Kalof. 1993. "Value Orientations, Gender, and Environmental Concern." *Environment and Behavior* 25 (5): 322–348. https://doi.org/10.1177/0013916593255002.
Whyte, Kyle, Chris Caldwell, and Marie Schaefer. 2018. "Indigenous Lessons about Sustainability Are Not Just for 'All Humanity'." In *Sustainability: Approaches to Environmental Justice and Social Power*, edited by Julie Sze, 149–179. New York University Press Scholarship Online, January 24, 2019.

Ecological Connectivity

Ecological connectivity (as defined by the UN Convention on Migratory Species) is the unimpeded movement of species, the connection of habitats without hindrance, and the flow of natural processes that sustain life on Earth (Liczner et al. 2024). Habitat loss and fragmentation are major threats to ecological connectivity (→ Habitat Loss). Connectivity is essential for many ecological processes, such as daily foraging, dispersal, gene flow, migration, species' abilities to shift ranges and adapt to the effects of climate change, and population persistence (→ Climate Change) (→ Biodiversity) (Liczner et al. 2024). Structural connectivity is the configuration of the landscape and habitat patches (size, shape, elevation), and functional (or behavioral) connectivity is the degree to which landscape facilitates or obstructs the movement of organisms between landscape patches (Fang et al. 2018). Connections across ecosystems, often at great distances, are a vital part of the functioning of the biosphere, but they can also be a major source of anthropogenic disruption including the introduction of invasive species (→ Telecoupling) (→ Invasive Animal Species).

Land-sea ecological connectivity is the interaction of physical, chemical, or biological processes between terrestrial and marine ecosystems, such as the migration and foraging of birds and fish in coastal areas and the transportation of nutrients from rivers (Fang et al. 2018). For example, the daily activities of birds enhance connectivity among different mangrove ecosystems in coastal areas, thus preventing the fragmentation of mangroves that are critical in land-sea ecological connectivity (Fang et al. 2018). Marine Protected Areas that incorporate connectivity and the maintenance of ecological corridors are essential in the dispersal of marine life across populations, communities, and ecosystems (Balbar and Metaxas 2018).

Wildlife corridors (also called linear habitats, wildlife movement corridors, biological corridors, green corridors, and dispersal corridors) are a human-made solution to the problem of urbanization and habitat fragmentation that enables the migration, colonization, and interbreeding of natural populations (→ Animal-Inclusive Design) (Clark et al. 2021). For example,

Canada's Banff National Park has a total of 44 wildlife crossing structures (six overpasses and 38 underpasses), and 82 km of highway fencing, the most crossing structures and enclosure fencing in a single location on the planet (Parks Canada 2022). A variety of large species have been recorded using Banff's wildlife crossings, including grizzly and black bears, wolves, coyotes, cougars, moose, elk, deer, bighorn sheep, wolverine, and lynx. Studies have confirmed that wildlife crossings facilitate gene flow for wildlife. Sawaya et al. (2014) found that 47 percent of black bears and 27 percent of grizzly bears who used crossings successfully bred, including both males and females of both species. Further, since the crossing and fencing structures were first constructed, wildlife-vehicle collisions in Banff National Park have dropped by more than 80 percent (→ Road Ecology). In addition, the Wallis Annenberg Wildlife Crossing in California (to be completed in late 2026) will connect habitat that has been fragmented by US Highway 101, one of the busiest freeways in the region, since 1926. The corridor will be the largest of its kind in the world (Santa Monica Mountains Conservancy 2026).

Corridors also save smaller animal species. A study of amphibian crossings in Monkton, Vermont, found that amphibian road mortality drops by over 90 percent with wildlife underpasses, showcasing their effectiveness in enhancing habitat connectivity (Marcelino et al., 2025). While corridors increase connectivity among fragments, they create edge effects (changes in climate, resources, or species, compared to the interior of a habitat) (→ Habitat Loss). Edge effects can negatively impact animal populations, such as increased predation of migrant birds and fish thus creating ecological traps for "edge-loving species," but mitigation is possible with wider corridors or the formation of softer edges that reduce the contrast between corridors (→ Ecological Traps) (→ Ecotones) (Haddad et al. 2014).

In a comparison of levels of biodiversity in continuous and fragmented landscapes, Gonçalves-Souza et al. (2025) conclude that fragmented landscapes consistently have lower biodiversity. Large, undisturbed landscapes are better for ensuring biodiversity than fragmented landscapes, and the restoration of habitat in large, continuous landscapes (such as one large forest) is essential to increase ecological connectivity and minimize biodiversity loss (→ Rewilding) (Gonçalves-Souza et al. 2025).

References

Balbar, Arieanna C. and Anna Metaxas. 2018. "The Current Application of Ecological Connectivity in the Design of Marine Protected Areas." *Global Ecology and Conservation* 17: e00569. https://doi.org/10.1016/j.gecco.2019.e00569.

Clark, Tanya, Tara Rava Zolnikov, and Frances Furio. 2021. "Urban Wildlife Corridors and Their Multiple Benefits." In *The Palgrave Encyclopedia of Urban and Regional Futures*, edited by Robert C. Brears, 1–4. Springer Nature. https://doi.org/10.1007/978-3-030-51812-7_32-1.

Fang, Xiaodong, Xiyong Hou, Xiaowei Li, Wan Hou, Masahiro Nakaoka, and Xiubo Yu. 2018. "Ecological Connectivity between Land and Sea: A Review." *Ecological Research* 33: 51–61.

Gonçalves-Souza, Thiago, Jonathan M. Chase, Nick M. Haddad, et al. 2025. "Species Turnover Does Not Rescue Biodiversity in Fragmented Landscapes." *Nature* 640: 702–706.

Haddad, Nick M., Lars A. Brudvig, Ellen I. Damschen, et al. 2014. "Potential Negative Ecological Effects of Corridors." *Conservation Biology* 28 (5): 1178–1187.

Liczner, Amanda R., Richard Pither, Joseph R. Bennett, et al. 2024. "Advances and Challenges in Ecological Connectivity Science." *Ecology and Evolution* 14: e70231. https://doi.org/10.1002/ece3.70231.

Marcelino, Matthew R., Steve G. Parren, and Brittany A. Mosher. 2025. "Assessing the Efficacy of Wildlife Underpasses in Mitigating Amphibian Road Mortality: A Case Study from the Northeastern United States." *Journal for Nature Conservation* 86: 126901. https://doi.org/10.1016/j.jnc.2025.126901.

Parks Canada. 2022. "Wildlife Crossing Structures and Research: Banff National Park." https://parks.canada.ca/pn-np/ab/banff/nature/conservation/transport/tch-rtc/passages-crossings.

Santa Monica Mountains Conservancy. 2026. "The Wallis Annenberg Wildlife Crossing at Liberty Canyon." https://smmc.ca.gov/liberty-canyon-wildlife-corridor/

Sawaya, Michael, Steven T. Kalinowski, and Anthony Clevenger. 2014. "Genetic Connectivity for Two Bear Species at Wildlife Crossing Structures in Banff National Park." *Proceedings of the Royal Society B: Biological Sciences* 281 (1780): 20131705. https://doi.org/10.1098/rspb.2013.1705

Ecological Footprint

The ecological footprint (EF) is a measure of the total impact of human actions on the biosphere (Global Footprint Network 2025). It was originally proposed in 1994 as a way to aggregate across specific kinds of impacts, such as land use change or energy use (Rees and Wackernagel 1994). The footprint calculations estimate how much land area would be required to support human activities in a sustainable way and compare it to the Earth's biologically productive area. The most recent estimate is that humans are using resources and generating impacts at a rate that would take 1.8 Earths to support sustainably. This excess is called "overshoot." Over the last 30 years, the EF has been applied to both individuals and nations (Rees and Wackernagel 2023). Debate about details of the methods used in the calculations continue, but the EF is widely used to demonstrate the substantial human impact on the biosphere and as a measure that can be used in research on what drives human impacts and how they might be reduced (York et al. 2003). Overall, the idea of the EF, over and above the specific calculations, is a useful tool for thinking about paths to sustainability.

Animal agriculture and fishing (including wild caught and farmed fish) are major components of the standard EF calculation, with each accounting for about 4 percent of the global EF or about 8 percent of total footprint for both combined (→ Livestock Production) (→ Aquaculture) (→ Insects as Food). Of course, in some countries the EF attributed to human consumption of animals for food can be much higher (→ Carnism). Calculations have been done of the "Ecological PawPrint" defined as the environmental impact of companion animals based on consumption and pet food waste (Martens et al. 2019). The EF of animals can be useful in analyzing ways to reduce stress placed on the environment. For example, the Ecological PawPrint approach has been used to examine the environmental impact (especially greenhouse gas emissions and land-use demands) of various alternative types of pet food (Alexander et al. 2020), and vegan pet foods tend to have the smallest impact (Nicholles and Knight 2025).

References

Alexander, Peter, Aiden Berri, Dominic Moran, David Reay, and Mark D.A. Rounsevell. 2020. "The Global Environmental Paw Print of Pet Food." *Global Environmental Change* 65 (November): 102153. https://doi.org/10.1016/j.gloenvcha.2020.102153.

Global Footprint Network. 2025. "Measure What You Treasure." Global Footprint Network. www.footprintnetwork.org/.

Martens, Pim, Bingtao Su, and Samantha Deblomme. 2019. "The Ecological Paw Print of Companion Dogs and Cats." *BioScience* 69 (6): 467–474. https://doi.org/10.1093/biosci/biz044.

Nicholles, Billy, and Andrew Knight. 2025. "The Environmental Sustainability of Meat-Based versus Vegan Pet Food." *Frontiers in Sustainable Food Systems* 9 (May): 1569372. https://doi.org/10.3389/fsufs.2025.1569372.

Rees, William E., and Mathis Wackernagel. 1994. "Ecological Footprints and Appropriated Carrying Capacity: Measuring the Natural Capital Requirements of the Human Capacity." In *Investing in Natural Capital: The Ecological Economics Approach to Sustainability*, edited by AnnMari Jansson, Monica Hammer, Carl Folke, and Robert Costanza. Island Press.

Rees, William E., and Mathis Wackernagel. 2023. "Ecological Footprint Accounting: Thirty Years and Still Gathering Steam." *Environment: Science and Policy for Sustainable Development* 65 (5): 5–18. https://doi.org/10.1080/00139157.2023.2225405.

York, Richard, Eugene A. Rosa, and Thomas Dietz. 2003. "Footprints on the Earth: The Environmental Consequences of Modernity." *American Sociological Review* 68 (2): 279–300.

Ecological Traps

Ecological traps form when animals are attracted to human-induced poor-quality habitats that compromise their survival. They are prevalent in both terrestrial and marine ecosystems and are usually caused by human modification of animal habitats (→ Habitat Loss). Ecological traps such as airports, agricultural fields, artificial wetlands, and urban areas threaten conservation efforts and could lead to population extinction (→ Extinction) (Battin 2004).

Most studies of ecological traps are of birds, and one of the best documented traps is the case of Cooper's hawks in Tucson, Arizona. Boal and Mannan (1999) compared the Cooper's hawks nesting ecology in the city with that of Cooper's hawks in rural areas in southeastern Arizona. Plentiful prey and nesting sites draw hawks to the city to nest, but nesting mortality was greater among urban compared to rural nests (more than 50 percent compared to less than five percent). The primary cause of death of nestlings was trichomoniasis, a disease carried by pigeons and doves who make up 84 percent of the urban hawk's diet, and the greatest cause of mortality among free-ranging Cooper's hawks was collisions with windows or vehicles.

Ecological traps also threaten the survival of marine species. For example, artificial reefs are used to enhance fisheries and mitigate impacts on natural reef habitat, but they are ecological traps for fish who inhabit them because they are more vulnerable to harvesting (→ Animal Forests) (Swearer et al. 2021). In addition, artificial reefs with large, sheltered areas attract big predatory fish, but without small shelters to protect prey fish the artificial reef becomes an ecological trap resulting in higher mortality among the prey fish (Swearer et al. 2021). Trap fishing is another example of an ecological trap for octopuses and lobsters because it exploits their inclination to shelter in holes or pots, thus increasing their mortality. The creation, modification, and restoration of habitats must consider whether the practices are unintentionally creating traps with negative outcomes for animals (→ Ecological Restoration) (Swearer et al. 2021).

The capacity for learning in animals can be used to avoid animal mortality from ecological traps (→ Learning for Survival). For example, train strikes are an ecological trap for wildlife and a leading cause of grizzly bear mortality in Banff National Park, Canada (→ Road Ecology) (→ Wilderness Animals). Bears are attracted to railways to feed on spilled grain and vegetation along the tracks. Since bears are rapid learners, they can be warned of approaching trains with simple warning devices such as ringing bells and flashing lights, mitigating their vulnerability to collisions (→ Coexistence) (St. Clair et al. 2019). Railway lines are also an ecological trap for deer who are attracted to the railway because of the need for iron in their diets: they lick the rails to pick up iron filings caused by the wheels grinding against the tracks. In Japan, train-deer collisions are avoided by fitting trains with a speaker that blasts the sound of a deer snorting to attract the animal's attention, followed by 20 seconds of dog barks to scare the deer off the tracks. Efforts to reduce ecological traps caused by anthropogenic activity are essential in the survival, reproductive success, and population viability of both terrestrial and ocean animals.

References

Battin, James. 2004. "When Good Animals Love Bad Habitats: Ecological Traps and the Conservation of Animal Populations." *Conservation Biology* 18 (6): 1482–1491.

Boal, Clint W., and R. William Mannan. 1999. "Comparative Breeding Ecology of Cooper's Hawks in Urban and Exurban Areas of Southeastern Arizona." *Journal of Wildlife Management* 63 (1): 77–84.
St. Clair, Colleen Cassady, Jonathan Backs, Alyssa Friesen, et al. 2019. "Animal Learning May Contribute to Both Problems and Solutions for Wildlife-Train Collisions." *Philosophical Transactions B* 374: 20180050.
Swearer, Stephen E., Rebecca I. Morris, Luke T. Barrett, Michael Sievers, Tim Dempster, and Robin Hale. 2021. "An Overview of Ecological Traps in Marine Ecosystems." *Frontiers in Ecology and the Environment* 19 (4): 234–242.

Ecology

The term ecology has two related meanings in common usage. The first is the interconnections among living organisms and the interconnections of living organisms with the non-living parts of the environment. We might speak of the ecology of a Colombian rainforest, the Sonoran Desert, or a suburban backyard. The second meaning of ecology is the scientific study of those interconnections, or the study of ecosystems (→ Ecosystems).

The term ecology was created by Ernst Haeckel in 1869 by fusing the Greek term "oikos," which means "household" with "ology," which means "the study of" (Worster 1977). Humans have always observed and thought about the interconnections of living organisms with each other and with the abiotic environment. Many other species clearly have cognitive understandings of these interconnections as well (→ Cognition). But systematic scientific study of ecology began in the nineteenth century, and by the twentieth century ecology became a major scientific discipline. Now there are professional societies, scientific journals, textbooks, and degree programs focused on the science of ecology (Begon and Townsend 2020).

Ecology has many subfields focusing on different kinds of ecosystems. For example, limnology is the study of lakes and rivers, and road ecology focuses on the impacts of human transportation structures on other species (→ Road Ecology). Subfields can also be defined in terms of the species studied (e.g., avian ecology, plant ecology). Human ecology examines the interaction between humans, other species, and the biophysical environment, drawing on both ecology and the social sciences. While there are many perspectives and theories in ecology, systems thinking and evolutionary perspectives influence nearly the entire field (→ Evolution). A systems perspective emphasizes the connections among the many parts of an ecosystem – how one species affects another and how aspects of the biophysical environment affect and are affected by living organisms. For example, many systems ecologists examine the flow of energy in ecosystems (Loft et al. 2026). The evolutionary perspective emphasizes that changes in ecosystems over time are the unfolding of both cultural and biological evolutionary processes, referred to as "the evolutionary play in the ecological theater" (Hutchinson 1965). Animals play a central role in that evolutionary play. Their actions can shape the structure of ecosystems – for example, when a keystone species is lost or added to an ecosystem, the structure of the system can change substantially (→ Keystone Species) (→ Ecosystem Engineers) (→ Rewilding) (→ Trophic Cascades). Animals who are mobile can cross ecosystem boundaries thus linking different ecosystems, including distant ecosystems (→ Ecotones) (→ Ecological Connectivity) (→ Telecoupling).

Sustainability requires the understanding and consideration of ecosystem dynamics, so ecology is central to sustainability analyses. Ecology is also a fundamental underpinning for other fields such as conservation biology and provides core scientific insights for many perspectives crucial to sustainability such as ecosystem services and coupled human and natural systems (→ Conservation Biology) (→ Ecosystem Services) (→ Coupled Human and Natural Systems). The insights of ecology are also central to many key dynamics that challenge sustainability, including commons dilemmas, extinction, and island biogeography (→ Commons Dilemma) (→ Extinction) (→ Island Biogeography).

References

Begon, Michael, and Colin R. Townsend. 2020. *Ecology: From Individuals to Ecosystems*. Wiley.

Hutchinson, G. Evelyn. 1965. *The Ecological Theater and the Evolutionary Play*. Yale University Press.

Loft, Ty, Imma Oliveras Menor, Nicola Stevens, et al. 2026. "Energy Flows Reveal Declining Ecosystem Functions by Animals across Africa." *Nature* 649: 104–112. https://doi.org/10.1038/s41586-025-09660-1.

Worster, Donald. 1977. *Nature's Economy*. Sierra Club Books.

Ecology of Fear

The concept of the ecology of fear explains "how predators can affect prey species' behavior, which in turn can influence other elements of the food web and produce effects of the same order of magnitude as those resulting from changes in predator or prey populations" (→ Trophic Cascades) (Ripple and Beschta 2004, 756). For example, prey species alter their foraging patterns and use of terrain according to the extent the terrain affects risk of predation. On landscapes with both open and closed habitats, ungulates may use a strategy of hiding in forest cover to reduce encounters with predators, or they may seek open terrain to see predators from a distance. Wolves are most successful at predation when they can closely approach prey without detection, thus ibex forage near "escape terrain" of extremely steep slopes or cliffs that are difficult for wolves and other predators to climb, and caribou move to elevations high above wolves who travel in valley bottoms (Ripple and Beschta 2004).

The presence of humans substantially impacts wildlife behaviors (→ Wilderness Animals). The closure of ecotourism with the onset of the pandemic in 2020 provided an opportunity for a natural experiment in the ecology of fear to test the effect of human presence on grizzly bear activity in the Khutze watershed on the central coast of British Columbia. Short et al. (2024) studied camera-trap recordings that detected grizzly bears before and after the closure of tourism. They found that daily bear detection rates decreased with human presence and increased when humans were absent. In addition, bears were detected more often at forested sheltered sites, while detection decreased at exposed sites when humans were present, probably due to the influence of habitat on bear perception of safety (Short et al. 2024).

The ecology of fear has also been studied in the foraging opportunities of mesopredators in anthropogenic environments (→ Mesopredator Release). Leighton et al. (2022) studied caracal (wild cat) feeding events at varying levels of urbanization in Cape Town, South Africa, and found that caracals living in the urban-dominated region selected feeding sites in resource-rich urban edges, despite associated risks of vehicle collisions, poaching, and domestic dog attacks, a behavior that leads the caracal into an ecological trap (→ Ecotones) (→ Ecological Traps). Caracals in the wildland-dominated region, however, avoided the urban edge, suggesting that the urban-living animals developed behavioral plasticity (such as using vegetative cover to "hide in plain sight") and habituation to human presence, most likely due to their frequent exposure to non-lethal encounters (Leighton et al. 2022).

Manipulative experiments have tested the ecology of fear in free-living wildlife and demonstrate that fear has the potential to not only kill prey (they die of fright) but also has substantial impact on prey fecundity. For example, in a study of fear of predators in free-living songbirds, researchers prevented all direct killing by predators at birds' nests and then manipulated fear by broadcasting predator calls and sounds for some nests, while other nests heard non-predator calls and sounds. The study found that parent birds who heard predators "laid fewer eggs, incubated them less well such that fewer hatched, and fed their offspring less often such that fewer survived, with the net result that they produced 40 percent fewer offspring" than parents who heard non-predator calls (Zanette and Clinchy 2019, R310).

Managing the ecology of fear can be a powerful tool to: 1) establish and maintain viable wild animal populations at a desired density; 2) change animal behavior and distribution to reduce conflict with people; 3) enhance people's positive experiences with wildlife including opportunities for hunting and outdoor recreation; and 4) eradicate invasive species or design reintroduction strategies for extirpated native species to increase or decrease predation and mortality rates for specific animal populations (→ Conflict with Terrestrial Animals) (→ Wildlife Tourism) (→ Invasive Animal Species) (→ Defaunation) (→ Rewilding) (Gaynor et al. 2021). The ecology of fear has consequences for population demography, species interactions, ecosystem functioning, and has potential applications to animal conservation and management challenges due to rapid global change (→ Global Environmental Change) (Gaynor et al. 2021).

References

Gaynor, Kaitlyn, Michael J. Cherry, Sophie L. Gilbert, et al. 2021. "An Applied Ecology of Fear Framework: Linking Theory to Conservation Practice." *Animal Conservation 24*: 308–321.

Leighton, Gabriella R.M., Jacqueline M. Bishop, Justin Merondun, Deborah J. Winterton, M. Justin O'Riain, and Laurel E.K. Serieys. 2022. "Hiding in Plain Sight: Risk Mitigation by a Cryptic Carnivore Foraging at the Urban Edge." *Animal Conservation* 25: 244–258.

Ripple, William J., and Robert L. Beschta. 2004. "Wolves and the Ecology of Fear: Can Predation Risk Structure Ecosystems?" *BioScience* 54 (8): 755–766.

Short, Monica L., Christina N. Service, Justin P. Suraci, Kyle A. Artelle, Kate A. Field, and Chris T. Darimont. 2024. "Ecology Fear Alters Behavior of Grizzly Bears Exposed to Bear-Viewing Ecotourism." *Ecology* 105: e4317. https://doi.org/10.1002/ecy.4317.

Zanette, Liana Y., and Michael Clinchy. 2019. "Ecology of Fear." *Current Biology* 29: R301–R316.

Ecosystem Engineers

Ecosystem engineers are animals (or plants) who directly or indirectly modulate the availability of resources (other than themselves) to other species by causing physical changes in the environment, thus modifying, maintaining, and creating habitats (Jones et al. 1994). Allogenic engineers (corals, sponges) (→ Animal Forests), change the environment via their own physical structures (their living and dead tissues). For example, reefs made by corals are direct sources of food and habitat for numerous organisms. Autogenic engineers (beavers, woodpeckers) change the environment by transforming living or non-living materials from one physical state to another (Jones et al. 1994). For example, woodpeckers transform wood without holes into wood with holes, indirectly providing resources for other creatures, such as nesting and roosting cavities for birds and bats (Jones et al. 1994).

In the marine environment, bivalves are important ecosystem engineers who produce habitat for other organisms (Meadows et al. 2012). Mussel beds act as a biological filter to help purify the water, and they enhance the abundance of other benthic species by providing refuge and enriched biodeposits (→ Animal Forests). Mussel shells provide attachment surfaces for other organisms, including seaweeds, insect larvae, and other invertebrates. The role of soil nematodes as ecosystem engineers is substantial – they may be the most important geobiological agent on the planet, affecting half of its surface (Meadows et al. 2012).

An ecosystem engineer can be a keystone species (→ Keystone Species), but the major difference is in the mechanism of impact on ecosystem structure. Keystone species are "outcome focused" with large effects on the ecosystem in spite of their low abundance (→ Trophic Cascades). Ecosystem engineers are "process focused" in directly shaping ecosystems by their actions (de Visser et al. 2013). For example, the beaver whose manipulations of waterways for their own benefit provide habitat for other species (outcome) in addition to cycling nutrients in the ecosystem (process). Whales are also both ecosystem engineers and a keystone species. As apex predators, whales have major trophic influences on marine ecosystems, and they also act as ecosystem engineers who influence species diversity and create physical changes to the ocean environment. Dead whales sink to the ocean floor where their carcasses transfer carbon from the surface to the deep sea (thus mitigating climate warming), yield massive amounts of organic enrichment to a nutrient-impoverished environment, provide habitat structure, and support diverse biotic assemblages. Because of their large impact, ecosystem engineers are central to natural processes critical to sustainability (Roman et al. 2014).

References

de Visser, Sanne, Elisa Thébault, and Peter C. de Ruiter. 2013. "Ecosystem Engineers, Keystone Species." In *Ecological Systems: Selected Entries from the Encyclopedia of Sustainability Science and Technology*, edited by Rik Leemans, 59–68. Springer.

Jones, Clive G., John H. Lawton, and Moshe Shachak. 1994. "Organisms as Ecosystem Engineers." *Oikos* 69: 373–386.

Meadows, Peter S., Azra Meadows, and John M.H. Murray. 2012. "Biological Modifiers of Marine Benthic Seascapes: Their Role as Ecosystem Engineers." *Geomorphology* 157–158: 31–48.

Roman, Joe, James A. Estes, Lyne Morissette, et al. 2014. "Whales as Marine Ecosystem Engineers." *Frontiers in Ecology and the Environment* 12 (7): 377–385.

Ecosystem Services

Ecosystem services are the many benefits humans receive from the way ecosystems function and the way species in them interact with each other including interactions with humans (→ Ecosystems) (Daily 1997). Some argue that the term "Nature's Contributions to People" (NCP) might be more useful than "ecosystem services" (Díaz et al. 2018). NCP is defined as "all the contributions, both positive and negative, of living nature (diversity of organisms, ecosystems, and their associated ecological and evolutionary processes) to people's quality of life" (Díaz et al. 2018, 270). It is still anthropocentric but efforts to broaden the concept are ongoing (→ Anthropocentrism). The motivation behind these efforts is to ensure broad consideration of ecosystems and their species in human decision-making.

Nature's contributions include provisioning humans with food, water, wood, fiber, and other material contributions. Other contributions are "regulating" – ecosystems moderate the flows of water, influence local weather, and have strong impacts on biogeochemical cycles, all of which influence global environmental change (→ Global Environmental Change). Nature's contributions can also be non-material, sometimes called "cultural ecosystem services" (Gould et al. 2025). These include appreciation of nature, the spiritual traditions of many cultures that consider nature and other species important, and the view that other species and functioning ecosystems have value in themselves over and above any human use of them (→ Ecocentrism).

In making decisions about ecosystems and sustainability, it is often easy to estimate a monetary value for provisioning contributions, such as the price of food products, wood, or water. It is much harder to find a monetary value for regulating non-material contributions, such as those that animals make to ecosystem stability. Economists and others have developed methods to estimate non-material contributions so that they are not ignored in making decisions about ecosystems (IPBES 2022). Some programs provide payments to governments and landowners who work to preserve and improve the flow of ecosystem services (→ Conservation Finance for Animals and Ecosystems). But the non-material contributions of animals to ecosystems are underrepresented, and efforts need to be made to more fully incorporate them (Chaplin-Kramer et al. 2025). For example, some countries not only recognize the value of animals to ecosystems, but provide distinct legal rights to non-human nature, including non-human animals, rivers, and ecosystems (→ Ecodemocracy) (→ Rights).

References

Chaplin-Kramer, Rebecca, Colleen R. Miller, Laura E. Dee, et al. 2025. "Wildlife's Contributions to People." *Nature Reviews Biodiversity* 1 (1): 68–81. https://doi.org/10.1038/s44358-024-00006-9.

Daily, Gretchen C. 1997. *Nature's Services: Societal Dependence on Natural Ecosystems*. Island Press.

Díaz, Sandra, Unai Pascual, Marie Stenseke, et al. 2018. "Assessing Nature's Contributions to People." *Science* 359 (6373): 270–272. https://doi.org/10.1126/science.aap8826.

Gould, Rachelle K., Terri Satterfield, Kirstin Leong, and Jonathan Fisk. 2025. "The Generations of Cultural Ecosystem Services Research." *Conservation Biology* 39 (5): e70065. https://doi.org/10.1111/cobi.70065.

IPBES. 2022. *Methodological Assessment of the Diverse Values and Valuation of Nature of the Intergovernmental Science-Policy Platform on Biodiversity and Ecosystem Services*, edited by Patricia Balvanera, Unai Pascual, Michael Christie, and David González-Jiménez. IPBES Secretariat. https://doi.org/10.5281/ZENODO.6522522.

Ecosystems

The term ecosystem was introduced by Arthur George Tansley to refer to all the living organisms and abiotic components assembled in a particular place (Tansley 1935; Pickett and Cadenasso 2002). "System" is intended to emphasize that there are interconnections among the organisms living in a place and between the organisms and the non-living components of the environment (→ Ecocentrism). At the largest level, the planet Earth (the biosphere) can be thought of as an ecosystem. The ecosystem concept can also be applied to very small systems, such as a puddle of water, the surface of a plant leaf, or a small rock outcropping in an ice field. The scale of the ecosystem is defined in terms of the analysis being done. All ecosystems are connected through flows of matter and energy and the movement of species, but boundaries between ecosystems can be defined by places where the biophysical environment and the mix of species shift (→ Ecological Connectivity) (→ Telecoupling) (→ Ecotones). An approach called landscape ecology sits between the analysis of the biosphere as a whole and individual ecosystems. It examines a landscape as an ecosystem defined by boundaries such as a large watershed (e.g., the Amazon) or other landform (e.g., the Sahara Desert) or by geopolitical boundaries.

Analyses of ecosystems differ in the extent to which they focus on the ecosystem as an integrated system versus as an assemblage of interacting species. The ecosystems ecology perspective emphasizes feedback mechanisms in ecosystems that keep the composition of species and biogeochemical processes stable over time and help ecosystems maintain their integrity even when stressed by outside changes like invasive species or global change (→ Invasive Animal Species) (→ Global Environmental Change) (Odum 1969). In this approach an ecosystem is seen as a "superorganism." Other approaches focus on change and instability and evolutionary processes (Corman et al. 2019). But all ecosystem perspectives can be deployed to understand ecosystem dynamics in support of efforts to move toward sustainability. For example, in some tropical forest ecosystems, overhunting has led to the loss of large vertebrates and that in turn substantially changes the structure and dynamics of the ecosystem (→ Defaunation) (→ Empty Forest) (→ Trophic Cascades). The core idea of an ecosystem (that all living things are interconnected with each other and with the biophysical environment) is at the center of thinking about sustainability and about the relationship between humans and other animals.

References

Corman, Jessica R., Scott L. Collins, Elizabeth M. Cook, et al. 2019. "Foundations and Frontiers of Ecosystem Science: Legacy of a Classic Paper (Odum 1969)." *Ecosystems* 22 (5): 1160–1172. https://doi.org/10.1007/s10021-018-0316-3.

Odum, Eugene P. 1969. "The Strategy of Ecosystem Development." *Science* 164: 262–270.

Pickett, Steward T.A., and Mary L. Cadenasso. 2002. "The Ecosystem as a Multidimensional Concept: Meaning, Model, and Metaphor." *Ecosystems* 5 (1): 1–10. https://doi.org/10.1007/s10021-001-0051-y.

Tansley, A.G. 1935. "The Use and Abuse of Vegetational Concepts and Terms." *Ecology* 16 (3): 284–307. https://doi.org/10.2307/1930070.

Ecotones

Ecotones are areas of transition between ecological communities, ecosystems, or ecological regions. Because of their small size and high diversity ecotones often can serve as centers of biodiversity and evolutionary novelty and can be a cost-effective target for conservation especially in the face of climate change (→ Biodiversity) (Kark 2013). The study of ecotones as a conceptual framework was developed in 1953 by Eugene Odum who argued that ecotones are not simply boundaries or edges, but rather in an ecotone there is active interaction between two or more ecosystems with properties that do not exist in either of the adjacent ecosystems (Kark 2013, 149; Odum 1953). Ecotones are dynamic. They support high levels of biological diversity; modulate flows of water, nutrients, and materials across the landscape; provide important components of wildlife habitat; and act as sensitive indicators of global environmental change (→ Global Environmental Change) (Risser 1995).

There are three types of ecotones: terrestrial (boundaries between different land ecosystems, such as forests and grasslands), aquatic (transitional areas between freshwater and marine environments, such as estuaries and wetlands), and riparian (areas along rivers and streams). Ecotones are classified as either natural (areas generated by environmental factors such as soils, geology, and climate) or anthropogenic (human-caused), such as forest clear-cut edges, margins between built-up and natural landscapes, and human-generated lakes and plantations.

Research on the effects of anthropogenic ecotones on biodiversity is often called "countryside biogeography" (Kark 2013). Some studies have found that ecotones provide valuable habitat and conservation opportunities for a variety of species from bush crickets (van der Mescht et al. 2021) to coyotes (Raymond and St. Clair 2023). In some ecosystems, natural and anthropogenic ecotones comprise different areas of transition for species. For example, in a study of different forest habitats for fruit-feeding butterflies, Lourenço et al. (2019) found that natural ecotones provide unique habitats for a variety of different butterfly species adapted to distinct ecological conditions, while anthropogenic ecotones were homogenous and supported only opportunistic butterfly species.

Coastal areas such as coral reefs in shallow waters, mangroves, salt marshes, and seagrass meadows are important natural ecotones connecting land and sea, and, due to human activity, they are some of the most highly threatened ecosystems on the planet (Levin et al. 2024). Non-habitat-forming animals (those who reside in the habitat but do not contribute to its structure) are often overlooked in coastal restoration efforts, but they can make critical contributions to the enhancement of ecosystem function, persistence, and resilience (→ Restoration Ecology). For example, bivalves reduce sulfide stress in seagrass habitats and increase the drought tolerance of saltmarsh vegetation (Sievers et al. 2022). Since most countries have coastal areas, it is important to understand how these ecotones are changing and what can be done to mitigate negative impacts on biodiversity and global environmental sustainability (Kark 2013).

References

Kark, Salit. 2013. "Effects of Ecotones on Biodiversity." *Ecological Systems: Selected Entries from the Encyclopedia of Sustainability Science and* Technology, edited by Rik Leemans, 147–160. Springer.

Levin, Noam, Rachel Madeline Cooper, and Salit Kark. 2024. "Quantifying Night Sky Brightness as a Stressor for Coastal Ecosystems in Moreton Bay, Queensland." *Remote Sensing* 16: 3828.

Lourenço Giselle M., Glória R. Soares, Talita P. Santos, et al. 2019. "Equal but Different: Natural Ecotones are Dissimilar to Anthropic Edges." *PLoS One* 14 (3): e0213008.

Odum, Eugene P. 1953. *Fundamentals of Ecology*. W.B. Saunders.

Raymond, Sage, and Colleen Cassady St. Clair. 2023. "Coyotes Access Diverse Anthropogenic Attractants at the Ecotone between Natural and Residential Urban Areas." *Urban Ecosystems* 26: 1589–1605.

Risser, Paul G. 1995. "The Status of the Science Examining Ecotones." *BioScience* 45 (5): 318–325.

Sievers, Michael, Christophr J. Brown, Christina A. Buelow, et al. 2022. "Greater Consideration of Animals Will Enhance Coastal Restoration Outcomes." *BioScience* 72 (11): 1088–1098.

van der Mescht, Aileen C., James S. Pryke, René Gaigher, and Michael J. Samways. 2021. "Ecological and Acoustic Responses of Bush Crickets to Anthropogenic and Natural Ecotones." *Biodiversity and Conservation* 30: 3859–3878.

Ecotourism

Ecotourism is defined as "travel with a primary interest in the natural history of a destination ... It is a non-invasive and participatory form of nature-based tourism that is built around learning, sustainability (conservation and local participation/benefits), and ethical planning, development and management" (Fennell 2020, 20). The core principles of ecotourism are sustainability, education, a nature base, and a conservation mandate or orientation. Ecotourism is an outlet for the innate tendency for humans to be attracted to other life forms and to affiliate with natural living systems (→ Biophilia) (Fennell 2020, 22). Ecotourists' experiences of nature firsthand should be non-invasive, placing the interests of the natural world, including individual entities, first, over the interests of the ecotourism industry (Fennell 2020).

Many kinds of tourism have been mislabeled as ecotourism, including ventures that are nothing more than visiting a natural area with no connection to conservation (e.g., nature-based tourism, consumptive wildlife tourism, adventure tourism, and outdoor recreation) (Stronza et al. 2019). Any activity that creates stress, pain, or causes death in animals should not be viewed as ecotourism, including all practices based on animal capture and confinement or other forms of animal use that cause suffering for human pleasure and entertainment (→ Wildlife Tourism) (Fennell 2013). Ecotourism as a practice ought to defend the interests of individual animals over the interests of humans, including encounters with free-living animals who would have the liberty to engage or terminate interactions independent of human influence (Fennell 2013).

Ecotourism has contributed to the survival of many threatened species, and at large scales, ecotourism can protect landscapes and entire wildlife populations (Stronza et al. 2019). For example, conservation efforts and a sustainable small-scale ecotourism operation have enabled an increase in the penguin colony in El Pedral, Patagonia, from six pairs in 2008 to 3200 pairs of penguins in 2020 (Garcia-Borboroglu and Reyes 2025). Argentina, New Zealand, Australia, and South Africa also have penguin colony protection areas where thousands of ecotourists visit annually under conditions of strict adherence to rules that minimize human disturbance to the colonies (Stronza et al. 2019; Desmond 2017).

In a landmark study assessing the influence of ecotourism on threatened species, Buckley et al. (2016) document that ecotourism supports threatened species through funding to public national parks; establishment of private reserves; switching communal land use from consumption to conservation; and funding of breeding, feeding, translocation, veterinary, anti-poaching, livestock-compensation, and conservation-incentive programs. On the other hand, threatened species can be impacted negatively by ecotourism through changes to plant and animal physiology, nutrition, pathology, reproduction, individual mortality, populations, and habitat (→ Habitat Loss). Significant effects of conservation through ecotourism have been documented for the orangutan and New Zealand sea lion. Different ecotourism intensities (zero/low, moderate, high) make the difference between eventual extinction and survival for the orangutan; for the sea lion, the net effect of ecotourism is the opposite – the impact of fisheries and ecotourism together increase pup mortality such that the population is projected to decline by approximately 45 percent over the next century (→ Defaunation) (Buckley et al. 2016, 7–8).

References

Buckley, Ralf, Clare Morrison, and J. Guy Castley. 2016. "Net Effects of Ecotourism on Threatened Species Survival." *PLoS One* 11 (2): e0147988.

Desmond, Jane. 2017. "Staging Privilege, Proximity, and 'Extreme Animal Tourism'." In *The Oxford Handbook of Animal Studies*, edited by Linda Kalof, 506–524. Oxford University Press.

Fennell, David. 2020. *Ecotourism*, 5th ed. Routledge.

Fennell, David A. 2013. "Ecotourism, Animals and Ecocentrism: A Re-examination of the Billfish Debate." *Tourism Recreation Research* 38 (2): 189–202.

Garcia-Borboroglu, Pablo, and Laura M. Reyes. 2025. "Conserving Penguins via Land and Sea Protection." In *Navigating Our Way to Solutions in Marine Conservation*, edited by Larry B. Crowder, 11–22. Open Book Publishers.

Stronza, Amanda L., Carter A. Hunt, and Lee A. Fitzgerald. 2019. "Ecotourism for Conservation?" *Annual Review of Environment and Resources* 44 (1): 229–253.

Empathy

Empathy is generally defined as the capacity to understand, feel, or respond to the emotional states of others, usually referring to humans. Scholars typically distinguish between two types of empathy: cognitive and affective (Eklund and Martina 2021; Whitley et al. 2026). Affective empathy is the ability to sense or physically experience the feelings or emotions of another (Maibom 2017). Cognitive empathy is the ability to understand the experience of others through imagination or perspective-taking (Spaulding 2017). Whitley and colleagues (2025, 3) expand the definition of empathy by Young et al. (2018) to be more inclusive of animals, where "empathy is an emotional state that relies on the ability to engage in one or more of the following: perception, understanding, feeling and/or action in response to the experiences or perspectives of another human or non-human animal." Scholarly inquiry on empathy toward the environment and the beings that inhabit it has emerged mostly within conservation psychology and ecocriticism. Conservation psychology focuses on how emotional and moral engagement with nature influences environmental concern and action, whereas ecocriticism examines how literary, cultural, and philosophical traditions shape understandings of ecological relationships and responsibilities (→ Conservation Psychology) (→ Ecocriticism). Together, these fields provide complementary perspectives on how empathy is cultivated, expressed, and interpreted in relation to the natural world.

Studies show that recognizing animals as sentient, emotionally expressive individuals can activate the same empathic processes that shape interpersonal behavior, prompting concern, moral reflection, and prosocial action (→ Sentience) (→ Cognition) (Spaulding 2017). In this sense, empathy extends beyond human relationships and becomes a central mechanism through which people interpret and evaluate their responsibilities toward other species. Empathy for animals has been linked to a range of ethical and behavioral outcomes. Individuals who perceive animals as cognitively aware and capable of suffering are more likely to support welfare reforms, reduce consumption of animal products, and engage in conservation efforts (→ Cognition) (→ Welfare) (→ Vegetarianism) (→ Veganism) (Young et al. 2018). Encounters with animals, whether through sanctuaries, zoos, or wildlife observation, frequently elicit empathic concern, particularly when animals exhibit recognizable emotions or vulnerabilities (→ Sanctuaries) (→ Zoos and Aquariums) (→ Wilderness Animals) (→ Wildlife Tourism) (→ Naturalist Programs). Research also demonstrates that empathy is not exclusively human; elephants console distressed herd members, rats free trapped companions, and some primates modify behavior in response to signs of fear or pain in others (Bartal et al. 2011; Plotnik and de Waal 2014). These findings challenge anthropocentric assumptions and highlight the evolutionary depth of empathy across species (→ Anthropocentrism). They also underscore why human empathy for animals is shaped not only by cognition, but by a broader social-emotional recognition of animals as relational actors (→ Cognition) (→ Agency) (→ Cultural Transmission).

The relationship between empathy and sustainability rests on the idea that emotional and moral regard for animals influences ecological decision-making (→ Ecocentrism) (→ Sustainability). Empathy can motivate individuals to adopt plant-based diets, support biodiversity protections, or choose coexisting-oriented strategies in human-wildlife conflict (→ Coexistence) (→ Conflict with Terrestrial Animals) (→ Conflict with Ocean Animals). At a policy level, acknowledging animals' emotional lives reinforces conservation approaches that value both species survival and individual well-being. By fostering deeper attention to interspecies relationships, empathy encourages forms of environmental action grounded in care, justice, and long-term ecological resilience (→ Environmental Justice) (→ Ecocentrism).

References

Bartal, Inbal Ben-Ami, Jean Decety, and Peggy Mason. 2011. "Empathy and Pro-Social Behavior in Rats." *Science* 334 (6061): 1427–1430.

Eklund, Jakob Håkansson, and Martina Summer Meranius. 2021. "Toward a Consensus on the Nature of Empathy: A Review of Reviews." *Patient Education and Counseling* 104 (2): 300–307.

Maibom, Heidi L. 2017. "Affective Empathy." In *The Routledge Handbook of Philosophy of Empathy*, edited by Heidi Lene Maibom, 22–32. Routledge.

Plotnik, Joshua M., and Frans B.M. de Waal. 2014. "Asian Elephants (*Elephas maximus*) Reassure Others in Distress." *PeerJ* 2: e278.

Spaulding, Shannon. 2017. "Cognitive Empathy." In *The Routledge Handbook of Philosophy of Empathy*, edited by Heidi Lene Maibom, 13–21. Routledge.

Whitley, Cameron T., Linda Kalof, L.C. Urquhart, et al. 2025. "Photo Portraiture Enhances Empathy for Birds with Potential Benefits for Conservation and Sustainability." *Sustainability* 17 (19): 8833.

Whitley, Cameron T., Kaitlin Barrailler, Mary Jackson, Theodore Bamberger, and Marta Burnet. 2026. "Empathy Toward Animals: Documenting Measurement Instruments Used in Research and Practice." *Journal of Zoological and Botanical Gardens* 7(2): 22.

Young, Ashley, Kathayoon A. Khalil, and Jim Wharton. 2018. "Empathy for Animals: A Review of the Existing Literature." *Curator: The Museum Journal* 61 (2): 327–343.

Empty Forest

Often called the "silent forest syndrome" because the sounds of wildlife are absent, an empty forest is an ecosystem that has not been completely degraded but is void of large mammals because of overhunting (Poulsen et al. 2023). Hunting has removed large animals from vast areas of neotropical forest where large, towering trees and native vegetation remain standing, but the ecological contributions provided by large vertebrates such as herbivory and seed consumption, seed dispersal, nutrient cycling, and predation are missing, (→ Trophic Cascades) (Redford 1992). For example, the human hunting of large mammalian and reptilian predators in neotropical aquatic ecosystems in Brazil, Colombia, and Peru has resulted in the removal of the largest carnivores (black caiman and giant otters) (→ Defaunation) (Redford 1992). As both apex predators and ecosystem engineers, black caimans are important in nutrient transfer from terrestrial to aquatic ecosystems (Redford 1992). The movement, burrowing, and dam and nest building activities of crocodilians such as caiman are critical in shaping the extent of water bodies, flow dynamics, the structure of shorelines, and habitat heterogeneity (→ Keystone Species) (→ Ecosystem Engineers) (He et al. 2024).

The loss of large-bodied animals from ecosystems, or body size downgrading, not only removes their ecological contributions but also increases the populations of herbivores and rodents and that in turn exacerbate ecosystem degradation (→ Mesopredator Release) (Poulsen et al. 2023). In a study of tropical mammal abundance and decline from hunting, Benítez-López et al. (2019) documented a decline of 13 percent across all mammal species, with medium-sized species reduced by more than 27 percent and large mammals by more than 40 percent. Future mammal populations are predicted to decline between 10 and 100 percent in approximately 50 percent of the pantropical forest area with declines greater than 70 percent in West Africa (Benítez-López et al. 2019). In heavily hunted forests in the Amazon, there have been combined population declines of up to 74.8 percent in 30 species of mid-sized mammals, birds, and reptiles, and the declines are even greater in the Atlantic Forest where mammal biomass declined more than 98 percent in intensively hunted sites (Bogoni et al. 2022). Because of rampant poaching for bushmeat and traditional medicine (→ Wildlife Trade), the forests of Vietnam and Laos have few animals heavier than 80 grams (0.18 pounds), and in Borneo's Lambir Hills National Park researchers couldn't find a single animal in 2007 who weighed more than a kilogram (Stokstad 2019).

While hunting-induced defaunation is now included in large-scale biodiversity assessments such as the Intergovernmental Science-Policy Platform on Biodiversity and Ecosystem Services (IPBES) and the Conference of the Parties to the Convention on Biological Diversity (CBD), interdisciplinary research strategies are needed to prevent further loss of fauna from tropical ecosystems (→ Convention on Biological Diversity). Krause and Tilker (2022) argue that sustainable forest management relies on the recognition that forests are social-ecological systems so mitigation requires holistic and place-based conservation measures, recognition of traditional ecological knowledge, and control of the commercial trade in tropical wildlife (→ Coupled Human and Natural Systems) (→ Indigenous and Decolonial Perspectives) (→ Wildlife Trade). For example, they note that the harvesting of wild meat and animal body parts from forest fauna can be a renewable resource if done sustainably. They argue defaunation must receive the same level of concern as that given to other global threats to humanity such as the global response to the SARS-CoV-2 pandemic (Krause and Tilker 2022). In the words of Redford (1992, 421), we "must not let a forest full of trees fool us into believing that all is well … an empty forest is a doomed forest."

References

Benítez-López, Ana, Luca Santini, Aafke M. Schipper, Michela Busana, and Mark A. J. Huijbregts. 2019. "Intact but Empty Forests? Patterns of Hunting-induced Mammal Defaunation in the Tropics." *PLoS Biology* 17 (5): e3000247.

Bogoni, Juliano A., Alexandre R. Percequillo, Katia M.P.M.B. Ferraz, and Carolos A. Peres. 2022. "The Empty Forest Three Decades Later: Lessons and Prospects." *Biotropica* 55 (1): 13–18.

He, Fengzhi, Jens-Christian Svenning, Xing Chen, et al. 2024. "Freshwater Megafauna Shape Ecosystems and Facilitate Restoration." *Biological Reviews* 99 (4): 1141–1163.

Krause, Torsten, and Andrew Tilker. 2022. "How the Loss of Forest Fauna Undermines the Achievement of the SDGs." *Ambio* 51: 103–113.

Poulsen, John R., Vincent Maicher, Halina Malinowski, and Camille DeSisto. 2023. "Situating Defaunation in an Operational Framework to Advance Biodiversity Conservation." *BioScience* 73: 721–727. https://doi.org/10.1093/biosci/biad079.

Redford, Kent H. 1992. "The Empty Forest." *BioScience* 42 (6): 412–422.

Stokstad, Erik. 2019. "The Empty Forest." *Science* 345 (6195): 397–399.

Endangered Species Act (ESA)

The Endangered Species Act (ESA) of 1973 (16 U.S.C. §§1531–1544) is a US federal law intended to prevent extinctions of plant and animal species (Sheikh and Ward 2021). Preventing species extinction is a fundamental aspect of most approaches to sustainability, and ESA and the Marine Mammal Protection Act (→ Marine Mammal Protection Act) are the key efforts of the US to prevent extinction. The major international effort to protect endangered species is the Convention on International Trade in Endangered Species (CITES) (→ Convention on International Trade in Endangered Species). Under the ESA scientific analysis is used to determine if a species is "endangered," i.e., is at risk of extinction in all or most of its range, or "threatened," i.e., at risk of becoming endangered in the foreseeable future (→ Conservation Biology) (→ Extinction). The decision to list a species as endangered or threatened is based on the best available science without consideration of economic consequences, although in late 2025 controversial proposals are being made to add economic considerations to the listing decision. By the end of 2025, 495 domestic animal species and 766 domestic plant species were listed by the US Fish and Wildlife Service as endangered and 249 animal species and 172 plant species as threatened (US Fish & Wildlife Service 2025).

Under the ESA it is illegal to harm a listed species. A major approach in protecting species and moving them toward recovery is designating areas as critical habitat. The critical habitat designation takes into consideration the economic costs of restricting human activities such as the grazing of cattle, logging, mining, and construction of buildings, dams, or highways in the habitat (→ Habitat Loss).

The ESA is considered one of the strongest extinction protection laws in the world (Eberhard et al. 2022; Schwartz et al. 2025). Some dramatic species recoveries are attributed, at least in part, to the ESA (Vlamis 2023). In 1973 there were about 1000 bald eagles in the US while there are now an estimated 300,000 individuals, and the number of grizzly bears in the lower 48 states has increased from less than 500 to 2000 individuals. California condors, gray wolves, humpback whales, peregrine falcons, and whooping cranes are also making strong recoveries. Overall, 55 species once listed have been "delisted" and are no longer considered endangered or threatened, and few animal or plant species have gone extinct in the US since the ESA was enacted (Schwartz et al. 2025).

The restrictions on human activity that come with protecting habitat and the costs and delays involved in obtaining permits for development have often led to complaints, lawsuits, and claims of substantial economic costs caused by the ESA. Perhaps the largest controversy involving many lawsuits and dramatic protests by both pro-logging and pro-protection activists occurred around efforts to limit logging in critical habitat for the spotted owl in the Pacific Northwest (Lawler 2023). But recent estimates suggest that overall, the economic impacts of the ESA are modest (Frank et al. 2025). Ongoing efforts try to find more effective, science-based strategies for protecting listed species, often working collaboratively with local communities and landowners (US Fish and Wildlife Service 2017). There are also concerns that species survival is necessary but not sufficient for sustainability and that policy should consider not just extinction but the loss of genetic diversity of a species, of species culture, and of the oldest animals in a population (→ Defaunation) (→ Cultural Transmission) (→ Learning for Survival).

References

Eberhard, Erich K., David S. Wilcove, and Andrew P. Dobson. 2022. "Too Few, Too Late: U.S. Endangered Species Act Undermined by Inaction and Inadequate Funding." *PLoS One* 17 (10): e0275322. https://doi.org/10.1371/journal.pone.0275322.

Frank, Eyal, Maximilian Auffhammer, David McLaughlin, Elisheba Spiller, and David Sunding. 2025. "The Cost of Species Protection: The Land Market Impacts of the Endangered Species Act." Working Paper W33352. Cambridge, MA: National Bureau of Economic Research. https://doi.org/10.3386/w33352.

Lawler, Katie. 2023. "What Happened in the Spotted Owl Controversy?" Bird Questions. November 3. Birdful.org. www.birdful.org/what-happened-in-the-spotted-owl-controversy/.

Schwartz, Mark W., Matthew A. Williamson, Joseph J. Apodaca, Alejandra Echeverri, Laura Melissa Guzman, and Kailin Kroetz. 2025. "The Fate of Imperiled Species: Lessons from 50 Years of the US Endangered Species Act." *Annual Review of Ecology, Evolution, and Systematics* 56 (1): 217–240. https://doi.org/10.1146/annurev-ecolsys-102723-064002.

Sheikh, Pervaze A., and Erin H. Ward. 2021. "The Endangered Species Act: Overview and Implementation." Report 46677, Library of Congress, Congressional Research Service. www.congress.gov/crs-product/R46677.

US Fish & Wildlife Service. 2017. "Safe Harbor Agreements." October 1. www.fws.gov/service/safe-harbor-agreements.
US Fish & Wildlife Service. 2025. "ECOSphere Environmental Conservation Online System." 20 July. https://ecos.fws.gov/ecp/report/boxscore.
Vlamis, Kelsey. 2023. "10 Iconic Animals That Were at Risk of Extinction but Rebounded after the Endangered Species Act Passed 50 Years Ago." *Business Insider*, November 25. www.businessinsider.com/animals-saved-endangered-species-act-passed-eagle-grizzly-wolf-alligator-2023-11.

Environmental Justice

Environmental justice refers to the fair distribution of environmental benefits and burdens and the meaningful involvement of all communities in decisions that shape their ecological conditions (Mohai et al. 2009). Although the term originally emerged from grassroots struggles against the disproportionate siting of hazardous waste facilities in predominantly Black and low-income neighborhoods (environmental racism), the idea of Black ecology is much older, dating back to the first Earth Day and start of the environmental movement (Hare 1970). This work has expanded into a broader framework concerned with how racism, class, and political marginalization structure exposure to environmental harms (Bullard 1993).

Contemporary approaches focus on critical environmental justice which expands traditional environmental justice to explain how systems of power such as racism, capitalism, colonialism, patriarchy, homophobia, state violence, and speciesism produce, and sustain environmental harm (→ Colonialism) (→ Speciesism) (Pellow 2025). Additionally, contemporary scholars argue that environmental justice encompasses forms of "slow violence" that are often invisible and extend far beyond documenting unequal exposures to pollution (Davies 2022). Critical environmental justice requires tracing the historical and structural conditions that produce disparities, acknowledging the expertise and lived experience of affected communities including Indigenous members, and advancing policies that safeguard the social and ecological well-being of both current and future generations (→ Indigenous and Decolonial Perspectives). This expanded definition underscores the reality that environmental problems are not simply technical failures, but social and political ones.

Environmental justice also encompasses non-human animals, especially those whose survival is closely tied to ecosystems disproportionately affected by industrial activity or environmental degradation (→ Habitat Loss) (Pellow 2013). For example, marine mammals such as orca and humpback whales depend on sound to navigate, forage, and communicate (→ Animal Languages). Industrial shipping lanes, military sonar, and offshore energy development introduce disruptive noise and chemical pollutants that interfere with these essential behaviors (Nabi et al. 2018). Many Indigenous and coastal communities whose identities, subsistence practices, and cultural relationships involve marine mammals experience profound losses when these animals suffer and die off. Legislation such as the Marine Animal Protection Act was designed and amended to address some of these issues (→ Marine Mammal Protection Act). Terrestrial landscapes reveal similar patterns. Research on intensive livestock production shows that pollution from these facilities, such as ammonia emissions, manure runoff, and antibiotic residues, harm the animals in them, the surrounding wildlife, and the predominantly rural, low-income, or minority communities who live nearby (→ Livestock Production) (Son et al. 2021). In each case, environmental injustices create intertwined vulnerabilities for human and non-human populations, where non-human animals serve as sentinels who provide early warning to the harmful effects of environmental degradation (→ Sentinels).

Integrating environmental justice into sustainability debates highlights the need for ecological solutions that address underlying social inequalities (Clark and Miles 2021). Sustainability initiatives that focus narrowly on technological innovations or conservation targets can inadvertently reproduce colonial or exclusionary practices that fail to consider community rights, local knowledge, or the everyday relationships people maintain with the animals and environments around them (→ Colonialism) (→ Indigenous and Decolonial Perspectives). Scholars note that conservation programs, especially those emphasizing land protection or species recovery, have occasionally displaced Indigenous communities or imposed top-down management strategies that ignore long-standing forms of multispecies stewardship (Pandey et al. 2025). Approaching sustainability through an environmental justice lens encourages models that respect both human and non-human agency, honors cultural relationships with place, and prioritizes equitable access to healthy ecosystems.

References

Bullard, Robert D. 1993. *Confronting Environmental Racism: Voices from the Grassroots*. South End Press.

Clark, Susan Spierre, and Monica Lynn Miles. 2021. "Assessing the Integration of Environmental Justice and Sustainability in Practice: A Review of the Literature." *Sustainability* 13 (20): 11238.

Davies, Thom. 2022. "Slow Violence and Toxic Geographies: 'Out of Sight' to Whom?" *Environment and Planning C: Politics and Space* 40 (2): 409–427.

Hare, Nathan. 1970. "Black Ecology." *The Black Scholar* 1 (6): 2–8.

Mohai, Paul, David Pellow, and J. Timmons Roberts. 2009. "Environmental Justice." *Annual Review of Environment and Resources* 34 (1): 405–430.

Nabi, Ghulam, Richard William McLaughlin, Yujiang Hao, et al. 2018. "The Possible Effects of Anthropogenic Acoustic Pollution on Marine Mammals' Reproduction: An Emerging Threat to Animal Extinction." *Environmental Science and Pollution Research* 25 (20): 19338–19345.

Pandey, Hari Prasad, Tek Narayan Maraseni, and Armando Apan. 2025. "Insights into Ecological Resettlements and Conservation-Led Displacements: A Systematic Review." *Environmental Management* 75 (5): 1281–1298.

Pellow, David N. 2013. "Environmental Justice, Animal Rights, and Total Liberation: From Conflict and Distance to Points of Common Focus." In *Routledge International Handbook of Green Criminology*, edited by Avi Brisman and Nigel South. Routledge.
Pellow, David Naguib. 2025. *What Is Critical Environmental Justice?* Wiley.
Son, Ji-Young, Marie Lynn Miranda, and Michelle L. Bell. 2021. "Exposure to Concentrated Animal Feeding Operations (CAFOs) and Risk of Mortality in North Carolina, USA." *Science of the Total Environment* 799: 149407.

Evolution

Life emerged on Earth a little less than four billion years ago (Westall 2025). In traditional evolutionary theory, animals are multicellular organisms who are heterotrophic (feed on others), usually mobile at some point in the life cycle, have flexible rather than rigid cell walls, and have complex nerve and muscle tissues. They diverged from single-celled forms of life called Choanoflagellates that could live singly or in colonies a little less than 1 billion years ago (Anderson et al. 2023). These and other changes in life on Earth are the result of evolution.

In Animal Studies and related areas of scholarship, there is a growing emphasis on an understanding that animals are part of a biotic community. Indeed, each individual animal is an ecosystem, the home for roughly as many bacterial and other cells as cells belonging to the host organism with the same number of viruses. For an adult human, that is about 10-100 trillion of each type of cell (Zilberg-Rosenberg and Rosenberg 2008; Sender et al. 2016). While it is convenient to think of individual animals, it is clear that evolution is acting on the whole ecosystem that is an individual within the larger ecosystem of other organisms, a "hologenome theory of evolution" (Zilber-Rosenberg and Rosenberg 2008). While most analyses must focus on part of the system, life on Earth is a series of ecosystems embedded in ecosystems from a single virus up to the biosphere (→ Ecocentrism) (Cañada et al. 2025).

Charles Darwin's (1859) *On the Origin of Species by Means of Natural Selection* introduced many key ideas of evolutionary theory. If a trait of an individual makes that individual more successful at reproducing, and the trait can be transmitted to other individuals, then that trait will become more prevalent in the population over time. Through this process the species adapts to the environment in which they live (→ Adaptation). Darwin's analysis was of traits that were transmitted through genetic inheritance. His (and subsequent analyses) noted that sometimes success in being transmitted to the next generation is a result of being better at finding food, avoiding predation, or other adaptations, a process called "natural selection." This process can lead to quite complex and elaborate adaptations to ecosystems, including interactions with other organisms. But organisms are also shaped by sexual selection in which traits that attract mates are favored even if they are costly in other ways. The peacock's tail is an example – the tail attracts mates but requires a great deal of physiological effort to produce. The theory of evolution has become the fundamental idea in biology and has proven to be one of the most flexible, profound and insightful ideas in science (Gould 2002).

The behavior of many species, including humans, is shaped by learning through observing others (→ Learning for Survival) (→ Cultural Transmission). Such behavioral traits are also subject to selection and lead to the evolution of culture (Boyd and Richerson 2024; Richerson and Boyd 2024). Thus, many species have a "dual inheritance" system in which both genetic and cultural evolution are underway. Evolutionary processes acting on culture will not always favor the same behavioral traits as those favored by selection acting on genes. For example, it is difficult for genetic evolution to produce altruism toward others who are not kin, but cultural evolution can often produce altruism (→ Altruism).

Evolution provides a powerful perspective for understanding biodiversity, how ecosystems function, and how life on Earth has changed over time (→ Biodiversity) (→ Ecosystems) (→ Extinction). It is an essential guide in developing effective strategies for sustainability. In particular, evolutionary analysis allows us to use the agency of humans and other species to guide our actions toward achieving improved well-being for all species while reducing stress on the environment, thus helping us to resolve commons dilemmas (→ Agency) (→ Commons Dilemma) (Richerson et al. 2024; Dietz 2005).

References

Anderson, Ross P., Christina R. Woltz, Nicholas J. Tosca, Susannah M. Porter, and Derek E.G. Briggs. 2023. "Fossilisation Processes and Our Reading of Animal Antiquity." *Trends in Ecology & Evolution* 38 (11): 1060–1071. https://doi.org/10.1016/j.tree.2023.05.014.

Boyd, Robert, and Peter J. Richerson. 2024. "Cultural Evolution: Where We Have Been and Where We Are Going (Maybe)." *Proceedings of the National Academy of Sciences* 121 (48): e2322879121. https://doi.org/10.1073/pnas.2322879121.

Cañada, Jose A., Salla Sariola, and Matthäus Rest. 2025. "Situating Microbes Within Complex Ecologies." *Engaging Science, Technology, and Society* 11 (1): 5–20. https://doi.org/10.17351/ests2023.2735.

Darwin, Charles. 1859. *On the Origin of Species by Means of Natural Selection, or the Preservation of Favoured Races in the Struggle for Life*. John Murray. http://darwin-online.org.uk/.

Dietz, Thomas. 2005. "The Darwinian Trope in the Drama of the Commons: Variations on Some Themes by the Ostroms." *Journal of Economic Behavior and Organization* 57 (2): 205–225.

Gould, Stephen Jay. 2002. *The Structure of Evolutionary Theory*. Harvard University Press.

Richerson, Peter J., and Robert T. Boyd. 2024. "Culture in Humans and Other Animals." *Science* 386 (6724): 846–847. https://doi.org/10.1126/science.adt8896.

Richerson, Peter J., Robert T. Boyd, and Charles Efferson. 2024. "Agentic Processes in Cultural Evolution: Relevance to Anthropocene Sustainability." *Philosophical Transactions of the Royal Society B: Biological Sciences* 379 (1893): 20220252. https://doi.org/10.1098/rstb.2022.0252.
Sender, Ron, Shai Fuchs, and Ron Milo. 2016. "Revised Estimates for the Number of Human and Bacteria Cells in the Body." *PLoS Biology* 14 (8): e1002533. https://doi.org/10.1371/journal.pbio.1002533.
Westall, Frances. 2025. "What the Earliest Evidence for Life Tells Us about the Early Evolution of the Biosphere." *Philosophical Transactions of the Royal Society B: Biological Sciences* 380 (1931): 20240106. https://doi.org/10.1098/rstb.2024.0106.
Zilber-Rosenberg, Ilana, and Eugene Rosenberg. 2008. "Role of Microorganisms in the Evolution of Animals and Plants: The Hologenome Theory of Evolution." *FEMS Microbiology Reviews* 32 (5): 723–735. https://doi.org/10.1111/j.1574-6976.2008.00123.x.

Extinction

Extinction occurs when the last member of a species dies. It also can occur when a species has changed enough through the process of evolution that paleontologists feel it should be classified as a new species and thus the ancestor to the species has disappeared (→ Evolution). Extinction is a normal part of the history of life on Earth. It has been estimated that over 99 percent of all species who have existed are now extinct (Raup 1992). As some species go extinct others arise through the typical processes of evolution.

The fossil record reveals five episodes of mass extinction in Earth's history (US National Park Service 2025; Ritchie 2022). The End Ordovician Extinction occurred 444 million years ago with 86 percent of species going extinct, probably because of extreme climate change toward a cold, highly glaciated planet (→ Global Environmental Change). The Late Devonian Extinction occurred about 372 million years ago, wiping out 75 percent of species. Many factors, including global climate change and changes in ocean temperature and chemistry, may have contributed to the event, but the relative importance of each is not known. The End Permian Extinction occurred about 252 million years ago, causing 96 percent of species to go extinct. Massive volcanic eruptions leading to climate change and ocean acidification seem to have been the cause. The End Triassic Extinction occurred about 200 million years ago, wiping out about 80 percent of species, and it is likely that volcanic eruptions produced climate change. The End Cretaceous Extinction occurred 66 million years ago because of the impact of an asteroid about 6-12 miles in diameter. This is the famous event that "wiped out" the non-avian dinosaurs and an estimated 76 percent of all species on Earth. Avian dinosaurs survived: birds of several lineages made it through, as did some non-bird reptiles, particularly crocodilians. Crocodilians are the closest living relatives of birds; with a 230-million-year history, they live today in the form of 26 crocodilian species, including crocodiles, caimans, and alligators (Kalof 2027). The history of extinctions over the last 65 million years or so is often used to estimate a background rate to which the recent rapid increase in extinctions can be compared (→ Applied Zooarchaeology).

Between 10,000-50,000 years ago many species of terrestrial megafauna (birds and mammals weighing more than about 100 pounds) went extinct, including the dire wolf, giant sloths, moas, mastodons, and saber-toothed cats. These extinctions coincide with the arrival of humans, with humans in the role of an invasive species (→ Invasive Animal Species) (Rowlands 2021; Lemoine et al. 2023). Human-caused global environmental change is now responsible for many species being at risk of extinction and a growing loss of biodiversity (→ Global Environmental Change) (→ Biodiversity). While hard to calculate and subject to debate, the rates may be high enough to be considered a "sixth great extinction" (Ceballos and Ehrlich 2023; Cowie et al. 2025; Wiens and Saban 2025). For example, the rate of terrestrial vertebrate extinction has been estimated at 8-100 times the background extinction rate (the standard or normal turnover of species) across evolutionary history and for invertebrates 500-1000 times the background rate. The Intergovernmental Science-Policy Platform on Biodiversity and Ecosystem Services estimates that approximately 1 million species are currently threatened with extinction, about 25 percent of all species examined, and the rate of extinction is accelerating (IPBES 2019). While work continues to more accurately assess which species are at risk and from what threats, the problem is widely acknowledged as a crisis. Protecting life on the land and in the oceans are two of the United Nations Sustainable Development Goals, and "Biosphere Integrity" is identified as one of the critical planetary boundaries that has been exceeded (→ Planetary Limits) (→ Sustainable Development Goals). CITES, the US Endangered Species Act, and many other efforts are working to protect biodiversity and prevent extinctions, and some species, such as the American alligator, the bald eagle, and the giant panda have moved from near extinction to viable populations (→ Convention on International Trade in Endangered Species) (→ Endangered Species Act).

References

Ceballos, Gerardo, and Paul R. Ehrlich. 2023. "Mutilation of the Tree of Life via Mass Extinction of Animal Genera." *Proceedings of the National Academy of Sciences* 120 (39): e2306987120. https://doi.org/10.1073/pnas.2306987120.

Cowie, Robert H., Philippe Bouchet, and Benoît Fontaine. 2025. "Denying That We May Be Experiencing the Start of the Sixth Mass Extinction Paves the Way for It to Happen." *Trends in Ecology & Evolution* 40 (8): 722–723. https://doi.org/10.1016/j.tree.2025.06.001.

IPBES. 2019. *Global Assessment Report on Biodiversity and Ecosystem Services of the Intergovernmental Science-Policy Platform on Biodiversity and Ecosystem Services.* IPBES Secretariat. https://ipbes.net/global-assessment.

Kalof, Linda. 2027. *Speaking of Real Animals and Their History: From the Deep Past to the Present.* Michigan State University Press.

Lemoine, Rhys Taylor, Robert Buitenwerf, and Jens-Christian Svenning. 2023. "Megafauna Extinctions in the Late-Quaternary Are Linked to Human Range Expansion, Not Climate Change." *Anthropocene* 44 (December): 100403. https://doi.org/10.1016/j.ancene.2023.100403.

Raup, David M. 1992. *Extinction: Bad Genes or Bad Luck?* Norton.

Ritchie, Hannah. 2022. "Did Humans Cause the Quaternary Megafauna Extinction?" *Our World in Data*, November 30. https://ourworldindata.org/quaternary-megafauna-extinction.

Rowlands, Mark. 2021. "The Great Dying." *World on Fire: Humans, Animals, and the Future of the Planet*, 157–171. Oxford Academic. https://doi.org/10.1093/oso/9780197541890.003.0008.

US National Park Service. 2025. "Mass Extinctions Through Geologic Time – Fossils and Paleontology." www.nps.gov/subjects/fossils/mass-extinctions-through-geologic-time.htm.

Wiens, John J., and Kristen E. Saban. 2025. "Recent Extinctions of Plant and Animal Genera Are Rare, Localized, and Decelerated." *PLoS Biology* 23 (9): e3003356. https://doi.org/10.1371/journal.pbio.3003356.

F

Feed Production

Feed production is the process of growing and transforming raw agricultural products and by-products into food for livestock and fish. Animal feeding in large, concentrated animal feeding operations (CAFOs), the predominant model for animal husbandry in the US, includes mixtures of feed grains, mill by-products, animal proteins, and micro ingredient formulations (vitamins, minerals, and antibiotics) (Sapkota et al. 2007). Specific feed ingredients vary by animal species, but there are concerns for human and non-human animal public health with the use of specific ingredients, such as rendered animal products, recycled animal waste, plant and animal fats, antibiotics, and heavy metals that result in higher levels of bacteria, antibiotic-resistant bacteria, prions, arsenic, and dioxin-like compounds in both animals and animal food products (Sapkota et al. 2007).

The production of feed for animals has a large environmental impact (→ Ecological Footprint). The production, processing, and transport of feed accounts for 45 percent of the greenhouse gas emissions from livestock, including 39 percent of methane emissions (→ Livestock Production). Feed-crop production occupies 33 percent of total arable land, 30 percent of all terrestrial land is given over to grazing, and over 90 percent of the global water use in the livestock sector is for irrigating feed crops (→ Feed Production) (Makkar 2018). Further, about 800 million tons of cereals (one-third of total cereal production) are used in animal feed, and this is projected to increase to over 1.1 billion tons by 2050 (Makkar 2018).

Global feed trade products are linked to about eight percent of livestock pressure on biodiversity, nine percent of deforestation, and 20 percent of the marine eutrophication impacts (Bidoglio et al. 2024). The global supply chains for agricultural products drive the displacement of environmental impacts from locations of consumption to the hotspots of feed production. For example, about 75 percent of global cultivated soybeans and 37 percent of cereals are allocated to feed livestock, and these geographically distant socioecological systems lead to food consumption occurring at a great distance from where the food is produced and the impacts occur (→ Telecoupling) (Bidoglio et al. 2024).

DOI: 10.4324/9781003528142-7

An alternative to traditional livestock feed that has the potential for making livestock production more sustainable is the use of novel resources as feed such as insects, leaf meals, spineless cactus, seaweed, and algae (→ Insects as Food) (Makkar 2018; Smetana et al. 2023). The use of novel feed ingredients in aquaculture could substantially reduce aquaculture's demand for forage fish, such as sardines, anchovies, and herring (→ Aquaculture). For example, replacing forage fish with microalgae, bacteria, yeast, and insects to feed farmed aquatic species reduces the use of forage fish that has been criticized for its impact on marine ecosystems (Cottrell et al. 2020).

The environmental impact of feed production from both livestock production and aquaculture can be reduced using a circular food system. Animals in circular food systems do not consume human-edible organic matter produced on arable land or harvested from water bodies, but are fed upcycled low-opportunity-cost food such as crop residues, by-products from food processing, food waste, and grass biomass (van Hal et al. 2023). A circular food system minimizes food loss and pollution and competes less for land and natural resources by recirculating products and materials, upcycling food by-products, and recycling nutrients back into the soil, leading to better soil health, biodiversity, water quality, and environmental health (Gatto et al. 2024).

References

Bidoglio, Giorgio, Florian Schwarzmueller, and Thomas Kastner. 2024. "A Global Multi-indicator Assessment of the Environmental Impact of Livestock Products." *Global Environmental Change* 87: 102853. https://doi.org/10.1016/j.gloenvcha.2024.102853.

Cottrell, Richard S., Julia L. Blanchard, Benjamin S. Halpern, Marc Metian, and Halley E. Froehlich. 2020. "Global Adoption of Novel Aquaculture Feeds Could Substantially Reduce Forage Fish Demand by 2030." *Nature Food* 1: 301–308. https://doi.org/10.1038/s43016-020-0078-x.

Gatto, Alessandro, Marijke Kuiper, Corina van Middelaar, and Hans van Meijl. 2024. "Unveiling the Economic and Environmental Impact of Policies to Promote Animal Feed for a Circular Food System." *Resources, Conservation & Recycling* 200: 107317. https://doi.org/10.1016/j.resconrec.2023.107317.

Makkar, H.P.S. 2018. "Review: Feed Demand Landscape and Implications of Food-Not Feed Strategy for Food Security and Climate Change." *Animal* 12 (8): 1744–1754.

Sapkota, Amy R., Lisa Y. Lefferts, Shawn McKenzie, and Polly Walker. 2007. "What Do We Feed to Food-Production Animals? A Review of Animal Feed Ingredients and Their Potential Impacts on Human Health." *Environmental Health Perspectives* 115 (5): 663–670.

Smetana, Sergiy, Anita Bhatia, Uday Batta, Nisrine Mouhrim, and Alberto Tonda. 2023. "Environmental Impact Potential of Insect Production Chains for Food and Feed in Europe." *Animal Frontiers* 13(4): 112–120.

van Hal, Ollie, Hannah H.E. van Zanten, Friederike Ziegler, Johan W. Schrama, Kiki Kuiper, and Imke J.M. de Boer. 2023. "The Role of Fisheries and Fish Farming in a Circular Food System in the European Union." *Sustainable Production and Consumption* 43: 113–123.

Feral Animals

Feral animals are domesticated species and their descendants who, after a history of living under the control of humans (→ Domesticated Animals), have returned to living in the wild as self-sustaining populations (Clutton-Brock 2012; Donaldson and Kymlicka 2011). The most common feral species are camels (in Australia), pigs (in Australia, the US, and South America), cats (worldwide), goats (in Australia and on various islands), and horses (in Australia, Western Europe, and the US). In urban environments, birds are common feral animals, including pigeons who live in towns and cities on every continent except Antarctica and parakeets who are familiar sights in US and UK cities (O'Connor 2013).

In the US, the best-known feral animal populations are the wild mustangs and burros who roam the Great Plains and Desert Southwest since their introduction by the Spanish in the fifteenth century. While horses first evolved in North America millions of years ago, they vanished from the continent around 11,000 BP primarily from climate change with human hunting also likely playing a role (Running Horse Collin et al. 2025). Feral horses have now successfully established populations on all continents except Antarctica (Lovász et al. 2025).

Feral herbivores can have substantial negative effects on ecosystems by altering plant structure and composition or by changing soil processes that in turn alter ecosystem function and biotic communities, thus reducing environmental quality (Eldridge et al. 2019). A meta-analysis of the global impact of feral horses on ecosystems documented a 13 percent reduction in environmental quality based on feral horse activity (Eldridge et al. 2019). However, in a study of the impact of feral horses on a desert environment, Ostermann-Kelm et al. (2009) found that while horse trailing (following on established trails) resulted in reduced vegetation cover and compacted soils, native plant diversity increased with the presence of horse feces and in areas adjacent to the trails. Soto et al. (2024) estimate a total cost of $141.95 billion worldwide from 18 invasive feral species with impacts mainly on the agricultural sector, and invasive feral species (livestock, pets, and beasts of burden) cost twice as much as their wild invasive species counterparts.

The negative ecological impacts of feral megafauna have generated conservation policies of eradication and culling, with "nativeness" being invoked as key to ecological health and using biodiversity data that count only native populations (→ Native Species) (→ Conservation Biology) (→ Conflict with Terrestrial Animals) (Lundgren et al. 2024). However, there is no evidence that feral megafauna have more negative impacts on plant diversity and abundance than native megafauna. Instead, functional traits (size and dietary preference) seem to shape impacts with larger-bodied and bulk-feeding megafauna promoting plant diversity (Lundgren et al. 2024).

In spite of their reputation for ecological damage, some argue that feral populations deserve conservation. Rodríguez-Rodríguez et al. (2022) identified three criteria essential in the biological conservation of feral animal populations: the conservation of genetic material, the preservation of ecological roles, and the importance of cultural heritage. Some examples of feral animals considered to be of conservation value are dromedary camels, Balearic wild goats, horses in Europe and North America, taurine cattle, dingoes, and the New Guinea singing dog.

Some countries are moving toward the protection of feral domestic animals (Bonacic et al. 2019) such as the Chilean law for responsible ownership that protects free-roaming dogs, the no-euthanasia policy of stray animals in Italy, the no-kill law of healthy animals in Poland, the anti-culling of street dogs in India, and the mandatory sterilization of stray dogs in the Netherlands (the first country to become "stray dog free"). In addition, feral species are granted legal rights or protected status as sentient beings in Ecuador, India, and Switzerland (→ Animal Law) (→ Ecodemocracy) (→ Rights) (→ Sentience) (Koumoundouros 2022). But international protection and support is lacking. For example, feral animals are excluded from the International Union for Conservation of Nature (IUCN) assessments (Rodríguez-Rodríguez et al. 2022). Granting rights to nonhuman animals, even those who are feral, follows an ecocentric perspective, a worldview that supports the value of all elements of nature as critical for global sustainability (→ Ecocentrism).

References

Bonacic, Cristián, Rocío Almuna, and J. Tomás Ibarra. 2019. "Biodiversity Conservation Requires Management of Feral Domestic Animals." *Trends in Ecology & Evolution* 34 (8): 683–686.

Clutton-Brock, Juliet. 2012. *Animals as Domesticates: A World View through History*. Michigan State University Press.

Donaldson, Sue, and Will Kymlicka. 2011. *Zoopolis: A Political Theory of Animal Rights*. Oxford University Press.

Eldridge, David J., Jingyi Ding, and Samantha K. Travers. 2019. "Feral Horse Activity Reduces Environmental Quality in Ecosystems Globally." Biological Conservation 241: 108367.

Koumoundouros, Tessa. 2022. "The First Country in the World has Given Legal Rights to Individual Wild Animals." *Science Alert: Nature*. www.sciencealert.com/the-first-country-in-the-world-has-given-legal-rights-to-individual-wild-animals.

Lovász, Lilla, Carolin Sommer-Trembo, Julia M.I. Barth, et al. 2025. "Rewilded Horses in European Nature Conservation: A Genetics, Ethics, and Welfare Perspective." *Biological Reviews* 100: 407–427.

Lundgren, Erick J., Juraj Bergman, Jonas Trepel, et al. 2024. "Functional Traits—Not Nativeness—Shape the Effects of Large Mammalian Herbivores on Plant Communities." *Science* 383: 531–537.

O'Connor, Terry. 2013. *Animals as Neighbors: The Past and Present of Commensal Species*. Michigan State University Press.

Ostermann-Kelm, Stacey D., Edward A. Atwill, Esther S. Rubin, et al. 2009. Impacts of Feral Horses on a Desert Environment. *BMC Ecology* 9 (22): 1–10. www.biomedcentral.com/1472-6785/9/22.

Rodríguez-Rodríguez, Eduardo J., Jesús Gil-Morión, and Juan J. Negro. 2022. "Feral Animal Populations: Separating Threats from Opportunities." *Land* 22: 1370.
Running Horse Collin, Yvette, Clément P. Bataille, Samantha Hershauer, et al. 2025. "Sustainability Insights from Late Pleistocene Climate Change and Horse Migration Patterns." *Science* 388 (6748): 748–755. https://doi.org/10.1126/science.adr2355.
Soto, Ismael, Paride Balzani, Francisco J. Oficialdegui, et al. 2024. "The Wild Cost of Invasive Feral Animals Worldwide." *Science of the Total Environment* 912: 169281.

Foundation Species

Foundation species are abundant, non-trophic (not related to feeding) animals (and plants) who have "a large effect on community structure by modifying environmental conditions, species interactions, and resource availability through their presence rather than their actions" (Wernberg et al. 2024, 248). Soulé et al. (2003) emphasize that foundation species are highly interactive, but because of their extreme abundance in an ecosystem, they cannot be classified as keystone species who have strong interactions disproportionate to their abundance (→ Keystone Species). A species is highly interactive when its absence leads to significant changes in the ecosystem, such as alteration in the flow of nutrients, loss of resilience to disturbance, and decreases in the diversity of native species (→ Native Species) (Soulé et al. 2003). The key difference between native species and foundation species is in origin: native species live naturally in a place, whereas foundation species (who can also be native) create physical conditions for an entire ecosystem.

Foundation species are often overlooked in conservation because they are common, but once they decline, their loss is hard to recover because of the myriad interactions they control (→ Defaunation). Their loss rapidly changes ecosystem function more than losing other species or interactions would (Ellison 2019). Some examples of foundation animal species include bison, prairie dogs, cod, bees, krill, mussels, and corals; foundation plant species include kelp, seagrass, mangrove wood, and cottonwood and aspen trees.

When some foundation species die, their structures provide substantial biogenic material in terrestrial, aquatic, and marine ecosystems, such as the organisms who make up dead shell mounds, reefs, and bioherms that span thousands of kilometers over the ocean floor (→ Animal Forests) (Saldaña et al. 2024). Because foundation species have traits similar to ecosystem engineers (high biomass, group-living, complex structures) (→ Ecosystem Engineers), they leave strong engineering legacy effects, or environmental modifications, that remain after death and persist as interacting components within and across environments – effects that can exceed or rival those of living foundational species (Saldaña et al. 2024).

Marine foundation species are particularly vulnerable to climate change, with impacts on genetic structures, physiology, and abundance, in addition to flow-on effects to associated communities, biodiversity, and ecosystem sustainability (→ Climate Change) (Wernberg et al. 2024). Recognition of the value of both living and dead foundation species in ecosystem functioning is essential in addressing the conservation challenges of the Anthropocene (→ Anthropocene).

References

Ellison, Aaron M. 2019. "Foundation Species, Non-trophic Interactions, and the Value of Being Common." *iScience* 13: 254–268. https://doi.org/:10.1016/j.isci.2019.02.020.

Saldaña, Patrick H., Christine Angelini, Mark D. Bertness, and Andrew H. Altieri. 2024. "Dead Foundation Species Drive Ecosystem Dynamics." *Trends in Ecology and Evolution* 39 (3): 294–305.

Soulé, Michael E., James A. Estes, Joel Berger, and Carlos Martinez Del Rio. 2003. "Ecological Effectiveness: Conservation Goals for Interactive Species." *Conservation Biology* 17 (5): 1238–1250.

Wernberg, Thomas, Mads S. Thomsen, Julia K. Baum, et al. 2024. "Impacts of Climate Change on Marine Foundation Species." *Annual Review of Marine Science* 16: 247–282.

G

Genetically Modified Organisms (GMOs)

Genetically Modified Organisms (GMOs) refer to animals and plants whose genome has been altered in ways that serve human ends (→ Anthropocentrism). Humans have been using selective breeding at least since the domestication of dogs and the preferential planting of seeds by food foragers (→ Artificial Selection and Breeding). The term GMO is reserved for organisms whose genome is altered by modern biotechnology that allows direct human intervention to alter the DNA and other parts of the genetic systems of a plant or animal (→ Biotechnology). In some cases, this is done by inserting genetic material from one organism into another, including the transfer of genes from plants or bacteria to animals (Fajardo et al. 2025).

Many applications of GMO technology are directed at animals used in agriculture and aquaculture (→ Livestock Production) (→ Feed Production) (→ Aquaculture) (Dolezel et al. 2025). These interventions are intended to improve the yields of the products desired by humans, improve the nutritional quality of animal products, increase disease resistance, and otherwise produce animals who are more useful to humans. However, relatively few countries have allowed GMO animals to become a routine part of the food system, and the EU prohibits any use of genetically modified animals for food. Beyond agriculture, GMO animals are being developed to produce medical products, such as organs that can be transplanted to humans (→ Animal Research). There are difficult issues in trying to assess the risks to human health of such organisms, including the possible benefits in terms of increased food production with lower environmental impact, socio-economic impacts, and the ethics of genetically modifying animals. Perhaps because the technology and its impacts are complex and unfamiliar, public opinion on GMOs and other biotechnologies is complex, although people are generally cautious (Azodi and Dietz 2019).

In contrast, GMO plants have been widely adopted in some countries. For example, 90 percent of the corn, cotton, and soybean crop areas in the US are planted with GMO varieties (→ Feed Production). These plants are designed to be pesticide resistant and disease resistant primarily to provide critical nutrients in human and animal diets, improve yields, and more productive. They are a continuation of the "green revolution" that bred food

DOI: 10.4324/9781003528142-8

plants to be more productive and to respond well to fertilizer. There are serious concerns about the impacts of GMO plants on sustainability, including increased commercialization of the agricultural system, risks to human health, and impacts on the environment and non-human animals (Noack et al. 2024). For example, it appears that adoption of GMO crops can have both positive and negative effects on bird populations, depending on both the species and the type of crop (Engist et al. 2024). The impact on bees seems similarly complex, with no evidence of toxicity from the GMO plants themselves, but the changes in management practices for herbicide-tolerant and insect-resistant crops may have indirect impacts on bee populations (Arpaia et al. 2021).

Some efforts to control insect-borne diseases, such as malaria and Lyme disease, introduce GMOs into ecosystems (→ Zoonosis). The GMO organisms have been modified so that they will slow transmission of the disease by reducing the population of the host (mosquitoes, ticks, etc.) or prevent it from carrying the disease. Such GMOs are often modified with "gene drives," genetic modifications that increase the spread of the engineered traits in the wild population. There are substantial ethical and ecological concerns with this technology that must be balanced with benefits to human and animal health (Moreno et al. 2024).

References

Arpaia, Salvatore, Guy Smagghe, and Jeremy B. Sweet. 2021. "Biosafety of Bee Pollinators in Genetically Modified Agro-ecosystems: Current Approach and Further Development in the EU." *Pest Management Science* 77 (6): 2659–2666. https://doi.org/10.1002/ps.6287.

Azodi, Christina B., and Thomas Dietz. 2019. "Perceptions of Emerging Biotechnologies." *Environmental Research Letters* 14 (11): 114018. https://doi.org/10.1088/1748-9326/ab4433.

Dolezel, Marion, Michael F. Eckerstorfer, Marianne Miklau, et al. 2025. "Governance Perspectives on Genetically Modified Animals for Agriculture and Aquaculture: Challenges for the Assessment of Environmental Risks and Broader Societal Concerns." *Animals* 15 (18): 2731. https://doi.org/10.3390/ani15182731.

Engist, Dennis, Laura Melissa Guzman, Ashley Larsen, Trevor Church, and Frederik Noack. 2024. "The Impact of Genetically Modified Crops on Bird Diversity." *Nature Sustainability* 7 (9): 1149–1159. https://doi.org/10.1038/s41893-024-01390-y.

Fajardo, Carlos, Marta Macedo, Tonka Buha, Marcos De Donato, Benjamin Costas, and Juan Miguel Mancera. 2025. "Genetically Modified Animal-Derived Products: From Regulations to Applications." *Animals* 15 (11): 1570. https://doi.org/10.3390/ani15111570.

Moreno, Ricardo D., Luca Valera, Cristián Borgoño, Juan Carlos Castilla, and José Luis Riveros. 2024. "Gene Drives, Mosquitoes, and Ecosystems: An Interdisciplinary Approach to Emerging Ethical Concerns." *Frontiers in Environmental Science* 11 (January): 1254219. https://doi.org/10.3389/fenvs.2023.1254219.

Noack, Frederik, Dennis Engist, Josephine Gantois, et al. 2024. "Environmental Impacts of Genetically Modified Crops." *Science* 385 (6712): eado9340. https://doi.org/10.1126/science.ado9340.

Global Environmental Change

Global environmental change (GEC) is large-scale change in the biosphere (Stern et al. 1992). Climate change is the most prominent example (→ Climate Change), but GEC also includes changes in the ozone layer, the hydrologic and other biogeochemical cycles, the large-scale distribution of plant and animal species by humans, land use and land cover, and the dispersal of long-lived toxic substances. While some of these processes occur at the local level across the planet (cumulative global changes), systemic GEC, such as climate change and changes in the ozone layer, operate at the planetary level (Turner, II et al. 1990). GEC impacts the abundance and distribution of animal species and the ecosystem services on which all species depend (→ Biodiversity) (→ Ecosystem Services).

GEC has occurred throughout Earth's history. About 717 million years ago, the concentration of greenhouse gases in the atmosphere dropped because of volcanic eruptions and the weathering of rocks. The average temperature fell to -50°C. Ice covered Earth from pole to pole. Life survived as microbes at the bottom of the frozen-over oceans. When a thaw came after about 85 million years, algae flourished leading to the first animals (Cutts 2025).

Starting at least with the spread of agriculture, humans have reached large enough population sizes and had sufficiently powerful technologies to drive cumulative global changes. Hunting and farming shift the distribution of plant and animal species. For the last 500 years, human action has also caused systemic global change. The massive deaths of humans in the Americas from colonial expansion and the spread of new infectious diseases disrupted Indigenous agriculture to the point where CO_2 absorbed by growth in abandoned fields contributed to a "little ice age" in the seventeenth and eighteenth centuries (→ Colonialism) (Koch et al. 2019). The Industrial Revolution and the massive increase in the use of fossil fuels such as coal and oil have produced climate change and ocean acidification. In the late twentieth century, the use of ozone-depleting chemicals such as chlorofluorocarbons (CFCs) led to reductions in the ozone layer in the stratosphere that protects the planet from dangerous ultraviolet radiation. International agreements have reduced the use of those chemicals, and the ozone layer is recovering, although it is now threatened by climate change (Chipperfield and Bekki 2024; Parson 2003).

Across Earth's history there have been mass extinctions (the loss of a high proportion of all species) at least five times (→ Extinction). The meteor impact that ended the dominance of dinosaurs is a dramatic example. Some argue that current GECs are driving a sixth mass extinction event, the first caused by humans (Ceballos and Ehrlich 2023), although others disagree (→ Extinction) (Wiens and Saban 2025). However, nearly all researchers concur that GEC and biodiversity loss are threats to the existence of many species and to sustainability. A recent international assessment concluded that the current extinction rate is "already at least tens to hundreds of times higher than it has averaged over the past 10 million years" (IPBES 2019, xvi). Overall, GECs driven by humans are large enough that some have argued Earth has entered a geological epoch dominated by humans, the Anthropocene (→ Anthropocene). The major dimensions of GEC have been used to define a set of "planetary boundaries" that many argue cannot be persistently exceeded if we are to achieve sustainability (→ Planetary Limits).

References

Ceballos, Gerardo, and Paul R. Ehrlich. 2023. "Mutilation of the Tree of Life via Mass Extinction of Animal Genera." *Proceedings of the National Academy of Sciences* 120 (39): e2306987120. https://doi.org/10.1073/pnas.2306987120.

Chipperfield, Martyn P., and Slimane Bekki. 2024. "Opinion: Stratospheric Ozone – Depletion, Recovery and New Challenges." *Atmospheric Chemistry and Physics* 24 (4): 2783–2802. https://doi.org/10.5194/acp-24-2783-2024.

Cutts, Elise. 2025. "Molecular Fossils Show How Life Survived Snowball Earth." *Science* July 17, 220–221.

IPBES. 2019. *Global Assessment Report on Biodiversity and Ecosystem Services of the Intergovernmental Science-Policy Platform on Biodiversity and Ecosystem Services*. Zenodo. https://doi.org/10.5281/zenodo.6417333.

Koch, Alexander, Chris Brierley, Mark M. Maslin, and Simon L. Lewis. 2019. "Earth System Impacts of the European Arrival and Great Dying in the Americas after 1492." *Quaternary Science Reviews* 207: 13–36.

Parson, Edward A. 2003. *Protecting the Ozone Layer: Science and Strategy*. Oxford University Press.

Stern, Paul C., Oran R. Young, and Daniel Druckman. 1992. *Global Environmental Change: Understanding the Human Dimensions*. National Academy Press.

Turner, B.L. II, Roger E. Kasperson, William B. Meyer, et al. 1990. "Two Types of Global Environmental Change: Definitional and Spatial-Scale Issues in Their Human Dimensions." *Global Environmental Change* 1 (1): 14–22. https://doi.org/10.1016/0959-3780(90)90004-S.

Wiens, John J., and Kristen E. Saban. 2025. "Recent Extinctions of Plant and Animal Genera Are Rare, Localized, and Decelerated." *PLoS Biology* 23 (9): e3003356. https://doi.org/10.1371/journal.pbio.3003356.

H

Habitat Loss

Habitat loss is a reduction in the amount of habitat for animal and plant species and is a major factor in the decline of global biodiversity (→ Biodiversity) (Banks-Leite et al. 2020). There are three main types of habitat loss: 1) destruction (extensive devastation of natural environments such as clear-cut logging), 2) degradation (changes to habitat quality without significant changes to its amount), and 3) fragmentation (conversion of large wild habitat areas into smaller ones) (Banks-Leite et al. 2020; Pons-Hernández 2024). Driven primarily by human activities, the loss of habitat endangers species' survival in both terrestrial (e.g., through land conversion for agriculture, mining, and urban development) and aquatic ecosystems (e.g., through dredging, pollution, and waste) (Pons-Hernández 2024). And all habitats on Earth are subject to degradation, fragmentation, and destruction as a result of climate change (→ Climate Change).

In a study of the global human impact on biodiversity in 97,783 affected sites, Keck et al. (2025) found that land-use change, resource exploitation, pollution, climate change, and invasive species all change the composition of biological communities. Habitat change and pollution (→ Pollution) had particularly strong effects on changes in the abundance and identity (composition) of biological communities and the diversity of terrestrial, freshwater, and marine ecosystems. There were significant differences in composition shifts between groups of organisms: microbes and fungi experienced the largest shifts, and mammals, fish, amphibians, and reptiles were less affected. These differences reflect the diversity, shorter life cycles, and higher dispersal rates of smaller species relative to larger species, and those characteristics in turn influence their response to human impact (Keck et al. 2025). They note that analyses of biodiversity must take account of spatial scale – loss of species in a small area may have different dynamics than loss of species across their range. In examining the drivers of local diversity loss, they again found that habitat loss and pollution were major drivers. However, in contrast to changes across a large spatial scale, at the local level the largest species experienced the strongest negative effects (Keck et al. 2025).

DOI: 10.4324/9781003528142-9

Amphibians are particularly vulnerable to habitat loss. They have low movement ability; high death rates when moving across roads and inhospitable terrain (→ Road Ecology); narrow habitat tolerance that exacerbates the effects of habitat loss, degradation, and edge effects; and high vulnerability to pathogens, invasive species, climate change, and environmental pollution (Cushman 2006). Roads restrict the movement of a wide variety of species in addition to amphibians, including bighorn sheep, bumblebees, European badgers, bobcats, coyotes, grizzly bears, ground beetles, roe deer, and turtles (→ Road Ecology) (Eigenbrod et al. 2008). The availability of "accessible habitat" (the amount of habitat that can be reached from a patch without crossing a road), reduces the effects of habitat loss and roads for species for whom roads are a major barrier to movement (Eigenbrod et al. 2008). Since flying wildlife maintain movement in fragmented landscapes and can have the same functions as non-flying wildlife, they can be a "biological insurance against the loss of ecosystem functions" provided by terrestrial wildlife whose movement is restricted (Zuluaga et al. 2022, 1).

Destruction, degradation, and fragmentation often have different effects, and those effects differ across species and can even vary for the same species in different parts of their geographic range (Banks-Leite et al. 2020; Riva et al. 2024; Watling et al. 2020). For example, tropical species are more negatively affected by habitat fragmentation because they have experienced relatively few habitat disturbances such as glaciations and fires over evolutionary time and thus have not evolved responses to disturbances. In contrast, temperate species have had to deal with disturbances over a long evolutionary history and are often more resilient to habitat change. Sometimes populations of the same species who live far from the range edge are either not affected or "can even be positively affected by habitat loss" (Banks-Leite et al. 2020, 673). These complexities present challenges for developing policies to protect biodiversity. However, current understanding indicates that prevention of habitat destruction, degradation, and fragmentation all matter, and that effective policies and programs must pay attention to the details of the ecosystems and species at risk (Riva et al. 2024).

Habitat loss also impacts interactions between species. Teckentrup et al. (2019) investigated small mammalian predator-prey interactions under simulated conditions of habitat loss. They found that perceived predation risk (fear) intensified the negative effects of habitat loss on prey populations (→ Ecology of Fear), with a loss of up to 30 percent in prey diversity and shifts in community composition from large to small animals, reinforcing the defaunation of large prey animals (→ Defaunation). The highest prey diversity was found in landscapes with non-fragmented safe patches that mitigated the negative effects of landscape change (→ Ecological Connectivity) (Teckentrup et al. 2019).

One approach to protecting species' habitat is to improve and preserve habitat in protected areas. The effective management of public lands for the protection and restoration of habitat is essential to any effective biodiversity strategy (Colburn 2007). Almost 15 percent of the Earth's land and 10 percent of its territorial waters are covered by national parks and other protected areas, and the coverage of marine protected areas increased by almost 300 percent between 2006 and 2016 (IUCN 2016). However, less than 20 percent of the world's key biodiversity areas are completely covered by protected areas. While the participants in the Convention on Biological Diversity have set a goal of protecting 30 percent of land area and 30 percent of marine area by 2030, progress has been slow (→ Convention on Biological Diversity).

Putting land and marine systems under protection does not guarantee an end to biodiversity loss. Eichenwald et al. (2020) studied the habitat loss for 24 vertebrates listed under the US Endangered Species Act and on the IUCN Red List across categories of land ownership (federal, state, private) in the US. The imperiled species studied lost the least amount of habitat (3.6 percent) on federal lands, whereas the amount of lost habitat on private lands was more than double (8.1 percent) (Eichenwald et al. 2020). In a review of the literature on the effectiveness of terrestrial protected areas, Geldmann et al. (2013) concluded that protected areas conserve forest habitat and are somewhat effective at maintaining species populations, but the connection between management actions (such as reducing poaching) and conservation outcomes is rarely evaluated, and the measurement of management actions is critical to adequately assess the protection of habitat in protected areas. Colburn (2007) argues that a restorative agenda and adaptive management for public lands can best be achieved through a deliberative process involving volunteers, citizens, and neighbors working together to protect and value nature for its own sake (→ Ecodemocracy) (→ Ecocentrism).

References

Banks-Leite, Cristina, Robert Mark Ewers, Hollie Folkard-Tapp, and Adam Fraser. 2020. "Countering the Effects of Habitat Loss, Fragmentation, and Degradation through Habitat Restoration." *One Earth* 3: 672–676.

Colburn, Jamison E. 2007. "Habitat and Humanity: Public Lands in the Age of Ecology." *Arizona State Law Journal* 39 (1): 145–208.

Cushman, Samuel A. 2006. "Effects of Habitat Loss and Fragmentation on Amphibians: A Review and Prospectus." *Biological Conservation* 128: 231–240.

Eichenwald, Adam J., Michael J. Evans, and Jacob W. Malcom. 2020. "US Imperiled Species are Most Vulnerable to Habitat Loss on Private Lands." *Frontiers in Ecology and the Environment* 18 (8): 438–446.

Eigenbrod, Felix, Stephen J. Hecnar, and Lenore Fahrig. 2008. "Accessible Habitat: An Improved Measure of the Effects of Habitat Loss and Roads on Wildlife Populations." *Landscape Ecology* 23: 159–168.

Geldmann, Jonas, Megan Barnes, Lauren Coad, et al. 2013. "Effectiveness of Terrestrial Protected Areas in Reducing Habitat Loss and Population Declines." *Biological Conservation* 161: 230–238.

IUCN. 2016. "The World Now Protects 15% of Its Lands, but Crucial Biodiversity Zones Left Out." September 3, 2016 [Press Release]. International Union for Conservation of Nature. https://iucn.org/news/secretariat/201609/world-now-protects-15-its-land-crucial-biodiversity-zones-left-out.

Keck, François, Tianna Peller, Roman Alther, et al. 2025. "The Global Human Impact on Biodiversity." *Nature* 641: 395–400. https://doi.org/10.1038/s41586-025-08752-2.

Pons-Hernández, Mònica. 2024. "Habitat Loss." In *Oxford Research Encyclopedia of Criminology and Criminal Justice*, edited by Henry N. Pontell. Oxford Academic. https://oxfordre.com/criminology/view/10.1093/acrefore/9780190264079.001.0001/acrefore-9780190264079-e-764.

Riva, Federico, Nicola Koper, and Lenore Fahrig. 2024. "Overcoming Confusion and Stigma in Habitat Fragmentation Research." *Biological Reviews* 99 (4): 1411–1424. https://doi.org/10.1111/brv.13073.

Teckentrup, Lisa, Stephanie Kramer-Schadt, and Florian Jeltsch. 2019. "The Risk of Ignoring Fear: Underestimating the Effects of Habitat Loss and Fragmentation on Biodiversity." *Landscape Ecology* 34: 2851–2868.

Watling, James I., Victor Arroyo-Rodríguez, Marion Pfeifer, et al. 2020. "Support for the Habitat Amount Hypothesis from a Global Synthesis of Species Density Studies." *Ecology Letters* 23 (4): 674–681. https://doi.org/10.1111/ele.13471.

Zuluaga, Santiago, Karina L. Speziale, and Sergio A. Lambertucci. 2022. "Flying Wildlife May Mask the Loss of Ecological Functions Due to Terrestrial Habitat Fragmentation." *Science of the Total Environment* 803: 150034.

I

Indigenous and Decolonial Perspectives

Indigenous perspectives are ways of knowing and relating to the world rooted in the traditions, practices, and worldviews of Indigenous people. Decolonial perspectives challenge dominant Western frameworks and seek to undo the harmful legacies of colonization by centering Indigenous knowledge, relationships, and sovereignty. In both cases animals are not resources to be managed, but relatives and community members in shared ecosystems. This idea of relational ontology challenges dominant Western paradigms that treat the environment and non-human animals as commodities (→ Anthropocentrism), but these perspectives encourage a move toward reciprocity and care. Globally, Indigenous peoples represent only around 5 percent of the current human population, but they have influence over the management of approximately 28 percent of Earth's terrestrial surface (Jiménez-Aceituno et al. 2025). Land and associated ecosystems managed by Indigenous communities fare better than those managed under other systems (Garnett et al. 2018; Jessen et al. 2022).

A growing number of Indigenous scholars define sustainability as the maintenance and renewal of reciprocal relationships among all beings including non-human animals. Robin Wall Kimmerer (Citizen Potawatomi Nation) (2013) asserts that Indigenous sustainability arises through reciprocity between humans and the more-than-human world. Leanne Betasamosake Simpson (Michi Saagiig Nishnaabeg scholar and member of the Alderville First Nation) (2017) extends this work emphasizing that reciprocity is continually renewed through relational accountability between humans, non-human animals, plants, land, and water, and that this ethic guides Indigenous resource management (→ Ecocentrism). Kyle Whyte (Citizen Potawatomi Nation) and colleagues (2018) offer that reciprocity and responsibility exist through extensive kinship networks not often recognized by Western culture, which contrasts with Western technocratic models of sustainability.

Some Indigenous and decolonial scholars argue that climate change, biodiversity loss, and other anthropogenic events are a continuation of colonial disruption and disasters that exist not only because of extreme events, but because of the erosion of reciprocal relationships between humans and ecosystems (→ Biodiversity) (→ Habitat Loss) (→ Colonialism) (→ Disasters). Along these

DOI: 10.4324/9781003528142-10

lines, Zoe Todd (Métis) (2018) writes about multispecies relations, demonstrating how Indigenous governance already contains sustainable practices that respect animals as partners rather than resources, and Kim TallBear (Sisseon Wahpeton Oyate) (2023) highlights how Indigenous traditions recognize animals as persons with agency and responsibilities, while Western culture treats animals primarily as resources (→ Agency) (→ Property). Decolonial approaches further highlight how colonial and capitalist systems disrupt animal relationships by making both land and species into commodities. Incorporating Indigenous and decolonial perspectives means recognizing that sustainable systems cannot be detached from social justice, cultural integrity, personhood, and land rights. For instance, in a 2025 study of 127 sustainability projects, scholars found that Indigenous-led or co-designed initiatives had a higher potential to shift value-systems and mindsets about the importance of sustainability (Jiménez-Aceituno et al. 2025).

Indigenous and decolonial perspectives fundamentally challenge Western paradigms that frame the environment and non-human animals as resources to be managed or consumed (→ Anthropocentrism). Instead, they articulate a relational ontology in which humans, animals, plants, and ecosystems exist within networks of reciprocity, responsibility, and mutual care (→ Ecocentrism). From this standpoint, sustainability is not a technical project of conservation or efficiency, but an ethical practice grounded in maintaining and renewing relationships among all living beings.

References

Garnett, Stephen T., Neil D. Burgess, Julia E. Fa, et al. 2018. "A Spatial Overview of the Global Importance of Indigenous Lands for Conservation." *Nature Sustainability* 1 (7): 369–374.

Jessen, Tyler D., Natalie C. Ban, Nicholas XEMŦOLTW Claxton, and Chris T. Darimont. 2022. "Contributions of Indigenous Knowledge to Ecological and Evolutionary Understanding." *Frontiers in Ecology and the Environment* 20 (2): 93–101.

Jiménez-Aceituno, Amanda, Aracely Burgos Ayala, E. Cepeda-Rodríguez, D.P.M. Lam, and Berta Martín-López. 2025. "Indigenous and Local Communities' Initiatives Have Transformative Potential to Guide Shifts toward Sustainability in South America." *Communications Earth & Environment* 6 (1): 481.

Kimmerer, Robin Wall. 2013. *Braiding Sweetgrass: Indigenous Wisdom, Scientific Knowledge and the Teachings of Plants*. Milkweed Editions.

Simpson, Leanne Betasamosake. 2017. *As We Have Always Done: Indigenous Freedom through Radical Resistance*. University of Minnesota Press.

TallBear, Kim. 2023. "Indigenous Genocide and Reanimation, Settler Apocalypse and Hope." *Aboriginal Policy Studies* 10 (2): 93–111.

Todd, Zoe. 2018. "Refracting the State through Human-Fish Relations: Fishing, Indigenous Legal Orders and Colonialism in North/Western Canada." *Decolonization: Indigeneity, Education & Society* 7 (1): 60–75.

Whyte, Kyle, Chris Caldwell, and Marie Schaefer. 2018. "Indigenous Lessons about Sustainability Are Not Just for 'All Humanity.'" In *Sustainability: Approaches to*

Environmental Justice and Social Power, edited by Julie Sze, 149–179. New York University Press Scholarship Online. https://doi.org/10.18574/nyu/9781479894567.003.0007.

Insects as Food

The consumption of insects (entomophagy) by both humans and other animals is a novel alternative protein food source that could lead to global food security and environmental sustainability. Insects have been a part of the human diet since the Pleistocene, and today more than two billion people worldwide regularly consume insects (Lange and Nakamura 2021). Sustainability advocates have recently become interested in edible insects because of their high nutritional value (quality protein, amino acids, and vitamins) and low environmental impact and because production requires less agricultural land and water and has fewer greenhouse gas emissions than conventional food production practices (Aguilar-Toalá et al. 2022).

Edible insects are classified into eight main orders: cockroaches and termites; beetles; flies; cicadas and stink bugs; bees, wasps, and ants; butterflies and moths; dragonflies; and crickets, grasshoppers, and locusts. Insects are cooked using traditional methods (boiling, roasting, stewing, baking, sun-drying) and sometimes processed through pasteurization, enzymatic proteolysis, and high-pressure methods (Liceaga 2022).

Some edible insects such as black soldier flies, yellow mealworms, and common house flies are important sources of food for animals, particularly fish, poultry, and swine (Sogari et al. 2019). The use of housefly larvae, yellow mealworms, and black soldier fly larvae meal in the diet of chickens results in equivalent weight and feed efficiency as a diet of fish, soybean, and corn meal (Ordoñez-Araque et al. 2022). Most of the industrial-scale edible insect farms are in Europe (particularly France and the Netherlands) and in North America (Canada and the US). They rear mainly crickets and yellow mealworms and do not use antibiotics, steroids, hormones, pesticides, or other synthetic chemical components that are often used in vertebrate livestock operations (Liceaga 2022).

Because of their short life cycle, insects can be reared on less land for short periods and achieve a higher intrinsic growth rate than traditional livestock with significantly less water and energy use (→ Livestock Production) (Liceaga 2022). Farm-raised insects require less resources than other domesticated animals to produce the same amount of protein – reared crickets, for example, use 161 square feet of land to produce 2.2 pounds of protein, while cattle use 2152 square feet (Liceaga 2022, 134). Further, livestock production emits 100 times more greenhouse gas emissions than mealworm production and 8–12 times more ammonia compared to cricket production (Liceaga 2022, 134).

In a review of the effect of edible insect supplementation on sustainability outcomes and health outcomes for both humans and other animals, Ros-Baró et al. (2022) found that human consumption of insects improved gut health, decreased systemic inflammation, increased blood concentrations of amino acids, and slowed digestion. Edible insect supplementation in the diet of non-human animals resulted in improved live weight, improved blood glucose and total cholesterol levels, and increased microbiota diversity. Environmental indicators (land use, water footprint, and greenhouse gas emissions) were 40–60 percent lower for the production of edible insects than traditional animal livestock (→ Ecological Footprint).

References

Aguilar-Toalá, José E., Rosy G. Cruz-Monterrosa, and Andrea M. Liceaga. 2022. "Beyond Human Nutrition of Edible Insects: Health Benefits and Safety Aspects." *Insects* 13 (1007). https://doi.org/10.3390/insects13111007.

Lange, Klaus W., and Yukiko Nakamura. 2021. "Edible Insects as Future Food: Chances and Challenges." *Journal of Future Foods* 1 (1): 38–46.

Liceaga, Andrea M. 2022. "Edible Insects, A Valuable Protein Source from Ancient to Modern Times." In *Emerging Sources and Applications of Alternative Proteins*, edited by Jianping Wu, 129–152. Academic Press. https://doi.org/10.1016/bs.afnr.2022.04.002.

Ordoñez-Araque, Roberto, Nadine Quishpillo-Miranda, and Luis Ramos-Guerrero. 2022 "Edible Insects for Humans and Animals: Nutritional Composition and an Option for Mitigating Environmental Damage." *Insects* 13 (944). https://doi.org/10.3390/insects13100944.

Ros-Baró, Marta, Patricia Casas-Agustench, Diana Alícia Díaz-Rizzolo, Laura Batlle-Bayer, Ferran Adrià-Acosta, Alícia Aguilar-Martínez, et al. 2022. "Edible Insect Consumption for Human and Planetary Health: A Systematic Review." *International Journal of Environmental Research and Public Health* 19 (18): 11653. https://doi.org/10.3390/ijerph191811653.

Sogari, Giovanni, Mario Amato, Ilaria Biasato, Silvana Chiesa, and Laura Gasco. 2019. "The Potential Role of Insects as Feed: A Multi-Perspective Review." *Animals* 9 (119).

Intersectional Environmentalism

Intersectional environmentalism has emerged as a contemporary analytic and political framework in environmental justice (→ Environmental Justice). Rooted in Black feminist and ecofeminist theory (→ Ecofeminism), it builds on Kimberlé Crenshaw's (1994) foundational articulation of intersectionality, that illuminated how overlapping structures of marginalization produce distinct and compounded forms of oppression. Intersectional environmentalism recognizes that oppression and social injustices are intertwined at multiple axes of discrimination (such as race, gender, and sexual orientation) and that these axes are connected to who is harmed and who is excluded from sustainable futures. It has gained broader recognition through the work of activist and writer Leah Thomas (2024).

An intersectional environmental lens is particularly important when considering the relationship between animals and sustainability. From such a perspective, non-human animals are embedded in the same systems of oppression that humans experience. These systems include industrial agriculture, colonial land dispossession, and imposed hierarchies of species (→ Livestock Production) (→ Colonialism) (→ Anthropocentrism). For instance, industrial animal agriculture or CAFOs (concentrated animal feeding operations) generate overlapping harms to underrepresented and marginalized human and non-human animal communities. These operations are disproportionately sited in low-income, rural, communities of color, concentrating air, and water pollution in already vulnerable communities while simultaneously subjecting thousands of non-human animals to intensive confinement and requiring underrepresented and marginalized individuals to perform repetitive, dangerous, and often traumatic tasks of animal slaughter and disassembly (Fitzgerald et al. 2009). Intersectional environmentalism reveals that these harms are co-produced by economic, racial, species-based, and environmental injustices. Critical environmental justice scholars like David Pellow (2021; 2014; 2017) assert that fights against pollution, dispossession, loss of species, and climate risk are also struggles against racial capitalism, ableism, and prison control. Indigenous and decolonial perspectives, widen the lens by centering land relations and kinship ethics, challenging anthropocentrism (→ Indigenous and Decolonial Perspectives).

Intersectional environmentalism demonstrates that environmental harm, racial capitalism, and species hierarchies operate through shared systems of oppression, making social and ecological justice inseparable. By emphasizing that non-human animals experience vulnerabilities alongside marginalized human communities in institutions such as industrial agriculture, it reframes sustainability as a fundamentally multispecies and relational endeavor (→ Multispecies Futures). Genuine sustainability requires dismantling the interconnected hierarchies that structure harm and advancing relationships of reciprocity, collective care, and social equity.

References

Crenshaw, Kimberlé Williams. 1994. "Mapping the Margins: Intersectionality, Identity Politics, and Violence Against Women of Color." In *The Public Nature of Private Violence: The Discovery of Domestic Abuse*, edited by Martha Albertson Fineman and Roxanne Mykitiuk, 93–118. Routledge.

Fitzgerald, Amy J., Linda Kalof, and Thomas Dietz. 2009. "Slaughterhouses and Increased Crime Rates: An Empirical Analysis of the Spillover from 'The Jungle' Into the Surrounding Community." *Organization & Environment* 22 (2): 158–184.

Pellow, David Naguib. 2014. *Total Liberation: The Power and Promise of Animal Rights and the Radical Earth Movement*. University of Minnesota Press.

Pellow, David Naguib. 2017. *What Is Critical Environmental Justice?* Wiley.

Pellow, David Naguib. 2021. "Black Lives Matter as an Environmental Justice Challenge." In *Lessons in Environmental Justice: From Civil Rights to Black Lives Matter and Idle No More*, edited by Michael Mascarenhas. Sage. https://doi.org/10.4135/9781544321974.n19.

Thomas, Leah. 2024. "Intersectional Environmentalism: The Sustainability Imperative Colloquium." *Advertising & Society Quarterly* 25 (2): 1–14. https://doi.org/10.1353/asr.2024.a932459.

Invasive Animal Species

An invasive animal species is an intentionally or unintentionally introduced organism who establishes a sustainable population in a region beyond the animal's native range and causes major ecological, health, or economic problems (Vitt and Caldwell 2014). Invasive species are a global problem that can lead to the extinction of some native species, threatening native biodiversity, the normal functioning of ecosystems, natural resources, regional economies, and human health (→ Native Species) (→ Extinction) (→ Ecosystems) (→ One Health) (Ricciardi 2013). The spread of invasive species has increased exponentially because of human activities such as international travel and global trade: tens of thousands of transoceanic ships transport several thousand non-native organisms on hulls and in ballast tanks on any given day (→ Telecoupling) (Ricciardi 2013). While most new species who enter an ecosystem have trouble establishing a niche and die out, those who are successful often have huge impacts. For example, as humans have migrated over the Earth, they have acted as invasive species, disrupting the ecosystems they enter and often leading to extinctions.

Among the most problematic invasive animal species on the planet (other than humans) are the spotted lanternfly who feeds on fruit, ornamental, and woody trees; feral domestic cats who kill billions of birds and mammals annually; the Burmese python who preys on mammals, birds, and reptiles; the European starling who competes with native bird species and destroys crops; and the red imported fire ant who impacts biodiversity, crops, and human health and whose range is increasing with the warming of the climate (Rao 2024).

Invasive species can profoundly affect ecosystems by altering community composition, resident species interactions, and habitat structure, and they can cause animal extinctions particularly in relatively closed systems with sharper ecotones such as lakes, river basins, and islands (→ Ecotones) (Ricciardi 2013). For example, on California's Channel Islands, the population of resident foxes had a decline almost to extinction with the arrival of the golden eagle who was historically an infrequent visitor to the islands (Roemer et al., 2002). When exotic pigs were introduced to the islands, they provided an abundant food source for the golden eagle who was then able to colonize. Once established the golden eagle preyed heavily on the island fox whose decline released populations of the endemic island skunk (→ Mesopredator Release) (Roemer et al., 2002).

In another example of how invasive species disrupt predator-prey dynamics, the introduction of the big-headed ant, one of the world's most widespread invaders, to the African savanna has disrupted the ability of native acacia ants to defend whistling-thorn trees from browsing by elephants, thus creating a landscape of higher visibility with less tree cover (Kamaru et al. 2024). Lions use tree cover to conceal themselves, and their success in hunting plains zebra is higher when visibility is lower. Big-headed ant invasion reduces the frequency of lion-zebra kills by changing the biophysical structure of the environment (→ Ecosystem Engineers) in ways that make trees vulnerable to elephant browsing thus increasing visibility across the landscape. As a result, lions have shifted their predation preference from zebras to buffalo.

Programs to eradicate invasive species need several conditions to be successful: the target species must be at low densities, with a biology susceptible to control, resources and public support must be in place, and re-invasion must be prevented (Ricciardi 2013). Such programs must also reflect biopolitical judgments about which forms of life are managed, suppressed, or protected (→ Biopolitics). Eradication can have unanticipated negative effects, particularly on islands with simple food webs. For example, the eradication of feral cats from Macquarie Island led to a population explosion of the European rabbit, an invasive herbivore (Ricciardi 2013). Given the scope of ecological damage by invasive species, the most effective management strategy for environmental sustainability is prevention, and invasions warrant as much attention in preparedness and response planning as natural disasters (→ Disasters) (Ricciardi 2013). Indigenous communities have often responded to invasive species in subtle and effective ways, and much could be learned from those practices (→ Indigenous and Decolonial Perspectives) (Reo et al. 2017).

References

Kamaru, Douglas N., Todd M. Palmer, Corinna Riginos, et al. 2024. "Disruption of an Ant-Plant Mutualism Shapes Interactions Between Lions and Their Primary Prey." *Science* 383: 433–438.

Rao, Devika. 2024. "5 of the Most Invasive Animal Species on the Planet." *The Week US Edition*. https://theweek.com/science/1026309/most-invasive-species.

Reo, Nicholas J., Kyle Whyte, Darren Ranco, Jodi Brandt, Emily Blackmer, and Braden Elliott. 2017. "Invasive Species, Indigenous Stewards, and Vulnerability Discourse." *The American Indian Quarterly* 41 (3): 201–223. https://doi.org/10.5250/amerindiquar.41.3.0201.

Ricciardi, Anthony. 2013. "Invasive Species." In *Ecological Systems: Selected Entries from the Encyclopedia of Sustainability Science and Technology*, edited by Rik Leemans, 161–178. Springer.

Roemer, Gary W., C. Josh Donlan, and Franck Courchamp. 2002. "Golden Eagles, Feral Pigs, and Insular Carnivores: How Exotic Species Turn Native Predators Into Prey." *Proceedings of the National Academy of Sciences* 99 (2): 791–796.

Vitt, Laurie J., and Janalee P. Caldwell. 2014. "Invasive and Exotic Species." In *Herpetology: An Introductory Biology of Amphibians and Reptiles*, edited by Laurie J. Vitt and Janalee P. Caldwell. Elsevier. https://doi.org/10.1016/C2010-0-67152-5.

Island Biogeography

Island biogeography examines the terrestrial, aquatic, and marine biodiversity of isolated natural continental and oceanic islands (→ Biodiversity). Islands that were once connected to the continental mainland have many species in common with the mainland; oceanic islands are small, reside on the oceanic crust, are mostly volcanic in origin, and have only those species who are able to disperse long distances over water (Helmus and Behm 2020). A third type of island is "artificial" and is made by draining water from seafloors or from dumping dredged sediment and sand onto seafloors and submerged coral reefs, processes that substantially impact marine species (Helmus and Behm 2020). But the theory applies to any type of habitat that occurs in patches in a larger landscape. For example, ponds, patches of forest in a savannah, parks in a city – all are "islands" from the perspective of the theory.

Island biogeography is determined by three fundamental processes of island species change: immigration, evolution, and extinction. Species immigrate to islands by dispersing (usually over water) with some species more likely to disperse than others (birds vs. frogs). Isolated from their source populations, island species evolve by adapting to island life and by randomly diverging from their source population; finally, species naturally go extinct from islands as a result of environmental change or random fluctuations in population size (Helmus and Behm 2020). The measure of biodiversity on an island is species richness (number of species), and the rates of immigration, evolution, and extinction are determined primarily by island area and isolation from source populations (Helmus and Behm 2020). The model describing these dynamics is considered a pathbreaking application of mathematics to ecology (→ Ecology) (MacArthur and Wilson 1967).

Humans have substantial impacts on immigration to islands through the introduction of nonnative species through global trade, thus expanding the potential species pool of immigrants who come from long distances (→ Invasive Animal Species) (→ Telecoupling). Nonnative species also drive evolution on islands, such as lizards who have evolved different behavior and morphology in response to the introduction of competitors and predators (Helmus and Behm 2020). Finally, over 60 percent of all known human-caused extinctions have occurred on islands, often due to invasive predators such as cats and rats, both of whom were the cause of the extinction of the dodo bird on the island of Mauritius by preying on dodo eggs and chicks (→ Invasive Animal Species) (→ Extinction) (Helmus and Behm 2020). Island environments are also impacted by the lack of predators in some ecosystems, such as on Lago Guri in Venezuela where herbivores have devastated the diversity and abundance of plant species in the absence of predator control (→ Mesopredator Release) (→ Trophic Cascades) (Terborgh et al. 2001).

Knowledge of island biogeography patterns for marine shallow-water organisms is critical for the conservation of marine systems. For example, the species richness of reef fishes, corals, marine snails, and lobsters are concentrated in island environments (Hachich et al. 2020). Marine shallow-water habitats have great biological and economic benefits: they shelter or are visited by 34 of the 35 extant animal phyla, half of whom are exclusively marine organisms, and they are important for coastline protection, food production, carbon dioxide sink, and nitrogen fixation (→ Animal Forests) (→ Ecological Connectivity) (Hachich et al. 2020).

References

Hachich, Nayara Fernanda, Débora Silva Ferrari, Juan Pablo Quimbayo, Hudson Tercio Pinheiro, and Sergio Ricardo Floeter. 2020. "Island Biography of Marine Shallow-Water Organisms." *Encyclopedia of the World's Biomes* 1: 61–75. Elsevier. https://doi.org/10.1016/B978-0-12-409548-9.11947-5.

Helmus, Matthew R., and Jocelyn E. Behm. 2020. "Island Biogeography Revisited." *Encyclopedia of the World's Biomes* 1: 51–56. Elsevier. https://doi.org/10.1016/B978-0-12-409548-9.11945-1.

MacArthur, Robert H., and Edward O. Wilson. 1967. *The Theory of Island Biogeography*. Princeton University Press.

Terborgh, John, Lawrence Lopez, Percy Nuñez V., et al. 2001. "Ecological Meltdown in Predator-Free Forest Fragments." *Science* 294: 1923–1926.

K

Keystone Species

Keystone species are animals (or plants) who have large effects on communities and ecosystems disproportionate to their abundance through numerous different processes such as trophic interactions, pollination, or habitat modification (→ Trophic Cascades). Keystone species shape the dynamics and structure of the ecosystem so strongly that if they are lost, substantial changes occur throughout the system. For example, wolves, dingoes, sea otters, and bees are all vital to the stability and health of their ecosystems despite their relatively small populations. Wolves help regulate herbivore populations and that in turn benefits the diversity of plant communities, and bees facilitate plant reproduction by transferring pollen that is necessary for the survival of many plant species. The term "keystone species" was coined by ecologist Robert Paine (1969) who performed a study of starfish along the rocky Pacific coastline. By removing the starfish from an area, the mussel population, the starfish's primary food source, exploded and overwhelmed the tidal pools, drastically decreasing the biodiversity in the area.

Keystone herbivores also have dramatic impacts on community structure. For example, Menge et al. (2013) note that the presence or absence of kangaroo rats in the Chihuahuan Desert determines whether the plant community is desert shrubland or arid grassland through the rats' seed predation and soil disturbance. Similarly, the bison's preference for competitively dominant grasses in the Konza Prairie in Kansas changes the composition, diversity, and physical and chemical aspects of grassland communities.

The concept of keystone shows how: 1) the loss of some species of low abundance may have surprisingly dramatic effects; 2) the preservation of a species of concern may depend on the distribution and abundance of other species with whom the target has no recognized interaction; and, conversely, 3) the loss of a species, such as a top carnivore, may reverberate to affect members of seemingly disparate species, such as plants or decomposers (Power et al. 1996).

Because keystone species are critical in the maintenance of species diversity and ecosystem functioning, their importance in ecosystem dynamics and

DOI: 10.4324/9781003528142-11

conservation is substantial. Caution is needed before management decisions are made that could result in the loss of a native species or the introduction of exotic species, and conservation should focus on ecosystems as well as species (→ Native Species) (→ Ecocentrism) (Menge et al. 2013).

References

Menge, Bruce A., Alison C. Iles, and Tess L. Friedenburg. 2013. "Keystone Species." In *Encyclopedia of Biodiversity*, 2nd ed., edited by Simon A. Levin, 442–457. Academic Press. http://dx.doi.org/10.1016/B978-0-12-384719-5.00081-2.

Paine, Robert T. 1969. "A Note on Trophic Complexity and Community Stability." *The American Naturalist* 103 (929): 91–93.

Power, Mary E., David Tilman, James A. Estes, et al. 1996. "Challenges in the Quest for Keystones." *BioScience* 46 (8): 609–620.

L

Learning for Survival

The ability to learn is so important to survival that all animals learn to some degree, even invertebrates who comprise 95 percent of the Earth's animal population learn to adapt to changing environments (Purdy 2012, 1910). For example, cephalopod mollusks (squid, cuttlefish, octopus, and nautilus) can signal each other, remember, and modify their behavior (Huffard 2013). Bees can learn a novel behavior (such as how to open a puzzle box to obtain food rewards) from a conspecific using a strategy they cannot carry out independently or learn through individual trial and error, indicating that they have a capacity for cumulative culture (→ Cultural Transmission) (→ Animal Cultures) (Bridges et al. 2024). In addition, recent research (Schleihauf et al. 2025) documents that nonhuman apes are capable of self-reflection, dismantling the belief that only humans are rational beings. Chimpanzees have the "spontaneous ability to weigh evidence of varying strengths, revise prior choices, and adapt when evidence is revealed to be unreliable" (Hare 2025, 445).

Behaviors learned from others persist longer and spread faster than asocial behaviors (learned from experience not social interaction). The cultural transmission of socially learned behaviors is common in many species, such as tool use among crows, sponge feeding in bottlenose dolphins, whale songs, and migratory patterns (→ Cultural Transmission) (Fugate et al. 2024). Migratory animals adapt to seasonal variability and environmental change through a variety of adjustments, including movement behaviors (→ Global Environmental Change) (Fugate et al. 2024). Ungulate migration is established and maintained by learning and the cultural transmission of information to conspecifics about how to optimally exploit their habitats: bison remember the locations of high-quality forage, white-tailed deer adopt the movement patterns of their mothers, and each generation of bighorn sheep and moose adds information from their own foraging experiences to the corpus of culturally transmitted information (Jesmer et al. 2018). Social animals who are reintroduced from captivity or translocated from the wild learn from same-species residents (Lewis et al. 2021). The protection of existing

DOI: 10.4324/9781003528142-12

migration corridors, stopover sites, and seasonal ranges are important not only to the existing landscape but also to the animals who depend on the traditional knowledge and culture that they need to sustain their populations (→ Ecological Connectivity) (Jesmer et al. 2018).

Social learning and cultural transmission are particularly important for young animals who must acquire knowledge from elders in the population. For example, young whooping cranes learn to migrate by following older individuals, and inexperienced animals follow older leaders' responses to changes or cues in the environment, as documented in pigeons, geese, and bats (Fugate et al. 2024). Longevity is one of the strongest predictors of stability in both terrestrial and aquatic species in part because the survival of older individuals buffers the population from environmental variability (Kopf et al. 2025). Old individuals in many animal populations have declined or been eliminated entirely because of human impacts. The oldest age class of fish has depleted in 79–97 percent of fish populations; old corals are being harvested; and poaching, trophy hunting, predator culling, and recreational harvesting for food often target the largest animals based on desired traits such as large antlers, horns, or tusks (→ Defaunation) (→ Wildlife Trade) (Kopf et al. 2025).

Among fish, socially learned knowledge is stored in the minds of older fish and learned by younger fish. But fishing activities reduce the abundance of older fish thus erasing the learned adaptations of long-lived individuals and leading to ecosystem-wide changes including disruptions in adaptation to prey, predators, and competitors (Wilson et al. 2024). The loss of older and larger fish has led to "a gradual transition in landings from long-lived, high trophic level, piscivorous bottom fish toward short-lived, low trophic level invertebrates and planktivorous pelagic fish" (Pauly et al. 1998, 860). Old, wise animals are vital to the maintenance of biodiversity, and the long-term sustainability of populations and ecosystems. Thus, longevity conservation should be management and restoration options for both ocean and terrestrial animals (Kopf et al. 2025).

Finally, in interspecies learning, individuals from different species observe, learn, and imitate each other's behaviors in patterns ranging from simple imitation to complex adaptations and problem-solving techniques (Sueur and Huffman 2024). In addition to the numerous cases of interspecies learning between humans and other species, migratory flycatchers learn habitat preferences from resident great tits, Himalayan black bears will follow on the ground below picking up acorns dropped by langur groups foraging in the trees above (Sueur and Huffman 2024), collaborative hunting occurs between octopus and fish, and ravens find wolf kills by relying on their memory of past wolf presence and wolf kill abundance (→ Multispecies Futures) (Loretto et al. 2026). The study of co-cultures is essential in understanding how all life is interconnected in the quest for environmental sustainability (→ Biocentrism).

References

Bridges, Alice D., Amanda Royka, Tara Wilson, et al. 2024. "Bumblebees Socially Learn Behaviour Too Complex to Innovate Alone." *Nature* 627: 572–578.

Fugate, Janey, Cody Wallace, Ellen O. Aikens, Brett Jesmer, and Matthew Kauffman. 2024. "Origin Stories: How Does Learned Migratory Behaviour Arise in Populations?" *Biological Reviews* 100 (2): 996–1014.

Hare, Brian. 2025. "Chimpanzees Are Natural Scientists." *Science* 390 (6772): 444–445. https://doi.org/10.1126/science.aeb7565.

Huffard, Christine L. 2013. "Social Learning in Cephalopods: A Natural History Perspective." In *Social Learning Theory: Phylogenetic Considerations Across Animal, Plant, and Microbial Taxa*, edited by Kevin B. Clark, 131–146. Nova Science.

Jesmer, Brett R., Jerod A. Merkle, Jacob R. Goheen, et al. 2018. "Is Ungulate Migration Culturally Transmitted? Evidence of Social Learning from Translocated Animals." *Science* 361 (6406): 1023–1025.

Kopf, R. Keller, Sam Banks, Lauren J.N. Brent, et al. 2025. "Loss of Earth's Old, Wise, and Large Animals." *Science* 387 (6729). https://doi.org/10.1126/science.ado2705.

Lewis, Mark A., William F. Fagan, Marie Auger- Méthé, et al. 2021. "Learning and Animal Movement." *Frontiers in Ecology and Evolution* 9: 681704. https://doi:10.3389/fevo.2021.681704.

Loretto, Matthias-Claudio, Kristina B. Beck, Douglas W. Smith, et al. 2026. "Ravens Anticipate Wolf Kill Sites Across Broad Scales." *Science* 391 (6790): 1151–1154.

Pauly, Daniel, Villy Christensen, Johanne Dalsgaard, Rainer Froese, and Francisco Torres, Jr. 1998. "Fishing Down the Food Web." *Science* 279: 860–863.

Purdy, Jesse E. 2012. "Learning in Invertebrates." In *Encyclopedia of the Sciences of Learning*, edited by N.M. Seel, 1910–1914. Springer. https://doi.org/10.1007/978-1-4419-1428-6_1043.

Schleihauf, Hanna, Emily M. Sanford, Bill D. Thompson, et al. 2025. "Chimpanzees Rationally Revise Their Beliefs." *Science* 390 (6772): 521–526. https://doi.org/10.1126/science.adq5229.

Sueur, Cédric, and Michael A. Huffman. 2024. "Co-cultures: Exploring Interspecies Culture Among Humans and Other Animals." *Trends in Ecology and Evolution* 39 (9): 821–829.

Wilson, James A., Jari Giske, and Clum Brown. 2024. "Overfishing Social Fish." *Fish and Fisheries* 26: 278–290.

Liminal Animals

Liminal animals are non-domesticated species, such as rodents, raccoons, squirrels, birds, and coyotes, who share space with humans without being entirely integrated into or wholly separate from human communities (Kalof and Whitley 2021). Often referred to as commensal species (living "together at table") liminal animals utilize modified or constructed human environments for shelter and sustenance, relying on these habitats either primarily or during critical periods in their life cycles that are essential for the viability of their populations (O'Connor 2013, 7).

Access to anthropogenic food sources drives most liminal animals into the human community, resulting in modifications in both physical and behavioral animal characteristics (Kalof and Whitley 2021). For example, skull size changes in Danish badgers and red foxes are linked to human food sources (Yom-Tov et al. 2003). Urban common hamsters in Russia show unique genetic features not seen in nonurban populations (Feoktistova et al. 2013), and urban coyotes in Canada suffer poor health from consuming low-quality, accessible human waste such as compost (Murray et al. 2015; 2016). Red foxes and coyotes are susceptible to sarcoptic mange, a disease that compromises their health (→ Zoonosis). As a preventative, some communities are treating the condition in foxes by injecting Ivermectin into meat baits at daily feeding stations of dry dog food over a 4–5-week period (Kalof and Whitley 2021). Food provisioning for wildlife such as supplemental feeding by tourists increases infection rates and changes host behavior and immunity, affecting wildlife exposure and tolerance to pathogens (→ Wilderness Animals) (Becker et al. 2015).

Many liminal animals are widely persecuted as pests and invaders of human spaces, and some species are vulnerable to ruthless extermination (Donaldson and Kymlicka 2011). People's attitudes toward liminal animals vary by culture, locality, and are often influenced by the size of the animal, such as coyotes who fall at the larger end of those able to adapt to human spaces, leading to more frequent human-animal conflicts (→ Conflict with Terrestrial Animals) (O'Connor 2013; Whitley et al. 2023). Hostility directed at coyotes and other mid-sized carnivores labeled as pests fuels modern wildlife killing contests, where thousands of coyotes, foxes, bobcats, and prairie dogs are killed with prizes going to the largest number of (or the largest animals) killed (Kalof and Whitley 2021). In the US, the Animal Damage Control Act has sanctioned the killing of liminal animals perceived to damage human property (→ Animal Damage Control Act). Humans have a long history of destroying liminal animals such as rats, mice, squirrels, sparrows, raccoons, coyotes, and foxes. The ritual public slaughter of agricultural pest animals was common in rural communities around ancient Rome, and bothersome animals such as flies, grasshoppers, locusts, and sparrows were excommunicated and exorcised during the animal trials of the Middle Ages (Kalof 2007).

Though often considered pests, the presence of liminal animals can benefit both the environment and humans. In Mumbai, leopards help lower stray dog populations, reducing dog bites and rabies risk (→ Zoonosis) (Braczkowski et al. 2018). Urban scavengers such as gulls, crows, and foxes quickly remove roadkill, an important ecosystem service (→ Ecosystem Services) (Kalof and Whitley 2021). In a simulated environment, Schwartz et al. (2018) found that seven species (mostly corvids) remove roadkill, and they conclude that roadkill removal by scavengers likely leads to underestimating roads' impacts on wildlife (→ Road Ecology). O'Bryan et al. (2018) argue for conservation strategies that will allow a variety of species to coexist in landscapes shared with humans (→ Coexistence). Making visible the benefits provided by "so-called problem animals" establishes tolerance in shared spaces and is an important step toward sustainability (O'Bryan et al. 2018, 229).

References

Becker, Daniel J., Daniel G. Streicker, and Sonia Altizer. 2015. "Linking Anthropogenic Resources to Wildlife–Pathogen Dynamics: A Review and Meta-Analysis." *Ecology Letters* 18: 483–495.

Braczkowski, Alexander R., Christopher J. O'Bryan, Martin J. Stringer, James E.M. Watson, Hugh P. Possingham, and Hawthorne L. Beyer. 2018. "Leopards Provide Public Health Benefits in Mumbai, India." *Frontiers in Ecology and the Environment* 16 (3): 176–182.

Donaldson, Sue, and Will Kymlicka. 2011. *Zoopolis: A Political Theory of Animal Rights*. Oxford University Press.

Feoktistova, Natalia Yu, Alexey Surov, Nikolay N. Tovpinetz, et al. 2013. "The Common Hamster as a Synurbist: A History of Settlement in European Cities." *Zoologica Poloniae* 58 (3–4): 116–129.

Kalof, Linda. 2007. *Looking at Animals in Human History*. Reaktion.

Kalof, Linda, and Cameron T. Whitley. 2021. "Animals in Environmental Sociology." In *International Handbook of Environmental Sociology*, edited by Beth Schaefer Caniglia, Andrew Jorgenson, Stephanie A. Malin, Lori Peek, David N. Pellow, and Xiaorui Huang, 289–314. Springer.

Murray, Maureen, Mark A. Edwards, Bill Abercrombie, and Colleen Cassady St. Clair. 2015. "Poor Health Is Associated with Use of Anthropogenic Resources in an Urban Carnivore." *Proceedings of the Royal Society B*: 282 (1806): 20150009. http://dx.doi.org/10.1098/rspb.2015.0009.

Murray, Maureen H., Jesse Hill, Peter Whyte, and Colleen Cassady St. Clair. 2016. "Urban Compost Attracts Coyotes, Contains Toxins, and May Promote Disease in Urban-Adapted Wildlife." *Ecohealth* 13 (2): 285–292.

O'Bryan, Christopher J., Alexander R. Braczkowski, Hawthorne L. Beyer, Neil H. Carter, James E.M. Watson, and Eve McDonald-Madden. 2018. "The Contribution of Predators and Scavengers to Human Well-being." *Nature Ecology & Evolution* 2: 229–236.

O'Connor, Terry. 2013. *Animals as Neighbors: The Past and Present of Commensal Species*. Michigan State University Press.

Schwartz, Amy L.W., Harry F. Williams, Elizabeth Chadwick, Robert J. Thomas, and Sarah E. Perkins. 2018. "Roadkill Scavenging Behavior in an Urban Environment." *Journal of Urban Ecology* 4 (1): 1–7.

Whitley, Cameron T., Melanie M. Bowers, and Harriett Grantz. 2023. "Urban Human–Coyote Conflicts: Assessing Friendliness as an Indicator of Coexistence." *Animals* 13 (18): 2903.

Yom-Tov, Yoram, Shlomith Yom-Tov, and Hans Baagøe. 2003. "Increase of Skull Size in the Red Fox (Vulpes vulpes) and Eurasian Badger (Meles meles) in Denmark during the Twentieth Century: An Effect of Improved Diet?" *Evolutionary Ecology Research* 5: 1037–1048.

Livestock Production

Livestock are domesticated terrestrial animals who are raised in controlled environments to provide a broad array of goods and services including food (meat, milk, eggs), energy (traction and transport), fibers, fertilizers, and hides (→ Domestic Animals) (Pandey and Upadhyay 2022). Livestock systems occur throughout the world. Those in developing countries contribute to the global production of 50 percent of beef, 41 percent of milk, 72 percent of lamb, 59 percent of pork, and 53 percent of poultry (Pandey and Upadhyay 2022). The demand for livestock products is stagnating in developed countries with increased production efficiency driven by technological developments in breeding, nutrition, and the use of transgenic livestock for food production may reduce environmental impacts (→ Biotechnology) (Pandey and Upadhyay 2022). In addition, while replacing meat from land mammals with poultry, fish, and aquatic invertebrates could reduce environmental impacts, the rising consumption of those lower-impact meat sources has not suppressed the international consumption of beef, mutton, and pig meat (York 2021).

In a review of research on the environmental impacts of different ruminant livestock on farm management systems, Kleppel (2020) found that regenerative-multi-paddock management (a nature-based practice focused on restoring soil health along with producing nutrient-dense food) reduces greenhouse gas emissions by more than 50 percent relative to industrial-conventional management, and regenerative management practices can remove from the atmosphere and store in the soil a significant portion of anthropogenic greenhouse gas emissions (→ Ecological Footprint). However, since economic optimization is the focus of modern industrial animal production, livestock are "finished" in confined animal feeding operations (CAFOs) that are widely used in the US, are increasingly limited in Europe, and are widely criticized as inhumane and unsafe (Kleppel 2020). Approximately half of the global livestock production is practiced using CAFOs. In the US, 97 percent of livestock production comes from CAFOs, and some companies have recently relocated their livestock production from Western Europe to the US to take advantage of weaker regulations (Fitzgerald 2019). The living conditions of animals in CAFOs are dire. Regardless of the species who must live in such confinement, whether cattle, buffalo, sheep, goats, pigs, or chickens, one thing is the same for all – "they are treated as machines which serves to implicitly deny their sentience" (→ Sentience) (Fitzgerald 2015, 32).

Livestock production requires significant resources. It occupies 30 percent of the world's land surface and 70 percent of all agricultural land surface, primarily in the cultivation of feed crops such as cereals and soybean used to feed cattle (→ Feed Production). Livestock production drives more than 8 percent of global water use (mainly irrigation for feed crops) (Hoque et al. 2022). Livestock production has a large environmental footprint as a contributor to 14.5 percent of the total greenhouse gas emissions that drive climate change (→ Ecological Footprint) (→ Climate Change) (Hoque et al. 2022). The global food system is also the primary driver of biodiversity loss because the conversion of natural ecosystems for crop production or pasture is the principal cause of habitat loss, with agriculture alone being a threat to 86 percent of species at risk of extinction globally (→ Habitat Loss) (→ Defaunation)

(→ Extinction) (Benton et al. 2021). In addition, climate change degrades habitats and causes species to disperse to new locations, bringing new species into contact and competition with each other, and creating opportunities for the emergence of infectious disease (→ Zoonosis) (Benton et al. 2021). In a study of the global impact intensities of five livestock categories, Bidoglio et al. (2024) found that the highest impact for both deforestation and biodiversity loss was from the production of ruminant meat (followed in order by eggs, pig meat, poultry meat, and milk), and for marine eutrophication the highest impact was from the production of pig meat (followed by poultry meat, eggs, ruminant meat, and milk).

Reform of our food system to protect biodiversity and the destruction of ecosystems and habitats depends on three strategies: 1) global dietary patterns need to be based more on plants, 2) more land needs to be protected and set aside for nature, and 3) farming needs to be more nature-friendly in a biodiversity-supporting way by replacing monoculture with polyculture farming practices (→ Vegetarianism) (→ Veganism) (→ Ecological Connectivity) (→ Restoration Ecology) (Benton et al. 2021).

References

Benton, Tim G., Carling Bieg, Helen Harwatt, Roshan Pudasaini, and Laura Wellesley. 2021. "Food System Impacts on Biodiversity Loss: Three Levers for Food System Transformation in Support of Nature." Research Paper, Energy, Environment and Resources Programme, Chatham House. www.chathamhouse.org/2021/02/food-system-impacts-biodiversity-loss.

Bidoglio, Giorgio, Florian Schwarzmueller, and Thomas Kastner. 2024. "A Global Multi-Indicator Assessment of the Environmental Impact of Livestock Products." *Global Environmental Change* 87: 102853. https://doi.org/10.1016/j.gloenvcha.2024.102853.

Fitzgerald, Amy J. 2015. *Animals as Food: (Re)connecting Production, Processing, Consumption, and Impacts*. Michigan State University Press.

Fitzgerald, Amy J. 2019. *Animal Advocacy and Environmentalism*. Polity Press.

Hoque, Mozammel, Akash Sukanta Mondal, and Satish Adusumilli. 2022. "Way Forward for Sustainable Livestock Sector." *In Emerging Issues in Climate Smart Livestock Production: Biological Tools and Techniques*, edited by Sukanta Mondal and Ram Lakhan Singh, 473–488. Academic Press.

Kleppel, Gary S. 2020. "Do Differences in Livestock Management Practices Influence Environmental Impacts?" *Frontiers in Sustainable Food Systems* 4 (141): 1–15.

Pandey, Hari Om, and Deepak Upadhyay. 2022. "Global Livestock Production Systems: Classification, Status, and Future Trends." *In Emerging Issues in Climate Smart Livestock Production: Tools and Techniques*, edited by Sukanta Mondal and Ram Lakhan Singh, 47–70. Academic Press.

York, Richard. 2021. "Poultry and Fish and Aquatic Invertebrates Have Not Displaced Other Meat Sources." *Nature Sustainability* 4 (9): 766–768. https://doi.org/10.1038/s41893-021-00714-6.

M

Marine Mammal Protection Act (MMPA)

The Marine Mammal Protection Act (MMPA) of 1972 (16 U.S.C. §§ 1361–1423h) is a US federal law to protect marine mammals (Congress 2022). With the US Endangered Species Act (16 U.S.C. §§ 1531–1544) and the Convention on International Trade in Endangered Species, the MMPA is a major tool to protect threatened and endangered species (→ Endangered Species Act) (→ Convention on International Trade in Endangered Species). The Act prohibits harassing, hunting, capturing, killing, importing, or exporting marine animals, and applies to dolphins, manatees, polar bears, sea lions, sea otters, seals, walruses, and whales in US waters. There are three federal agencies that manage the MMPA: the National Oceanic and Atmospheric Administration (NOAA), the US Fish and Wildlife Service (FWS), and the Marine Mammal Commission (MMC). NOAA is primarily responsible for monitoring marine mammals in US oceans and in beached or stranding situations. The FWS focuses on marine mammals farther inland and those in freshwater ecosystems. The MMC is designed to operate as an agency providing oversight and recommendations for other agencies. The goal of the MMPA is to prevent the unnecessary depletion or extinction of marine mammals, and to keep an "optimum sustainable population," that allows for species to continue to contribute to ecological systems and not just avoid extinction (→ Defaunation) (→ Extinction).

The MMPA developed out of the environmental movement in the late 1960s and early 1970s when scientists and the public became alarmed about the large number of dolphins killed in tuna fisheries and the failure of international bodies to control commercial whaling (Roman et al. 2013). The US Congress responded by creating a precautionary law that prioritized ecosystem health over short-term harvest. Over time, the Act has been amended to address new issues, such as bycatch reduction plans, human-animal conflicts, stock assessment reports for marine mammal populations, and co-management with Alaska Native communities who hunt marine mammals for subsistence (→ Indigenous and Decolonial Perspectives) (→ Conflict with Ocean Animals) (Roman et al. 2013; Moore et al. 2021). Currently, the focus of the MMPA is on regulating incidental take in commercial fisheries, limiting ship strikes and

DOI: 10.4324/9781003528142-13

underwater noise (Southall et al. 2021), and guiding conservation planning for depleted or threatened population.

The MMPA is closely tied to sustainability because it links species protection to the long-term health and resilience of ocean ecosystems. By requiring that human activities do not push marine mammals below sustainable population levels, the Act helps maintain biodiversity, food-web stability, and the ecological roles of top predators and other key species (→ Biodiversity) (→ Keystone Species) (→ Trophic Cascades). Healthy marine mammal populations support sustainable coastal economies through fisheries, tourism, and cultural practices, while also signaling broader ocean health in a warming world (Nelms et al. 2021). Conservation organizations and scientists frequently point to the MMPA as an early example of ecosystem-based, precautionary management that aims to balance human use of marine animals.

References

US Congress. 2022. *Marine Mammal Protection Act of 1972*. US Government Publishing Office. www.govinfo.gov/app/details/COMPS-1679.

Moore, Jeffrey E., Dennis Heinemann, Tessa B. Francis, et al. 2021. "Estimating Bycatch Mortality for Marine Mammals: Concepts and Best Practices." *Frontiers in Marine Science* 8: 752356.

Nelms, Sarah E., Joanna Alfaro-Shigueto, John P.Y. Arnould, et al. 2021. "Marine Mammal Conservation: Over the Horizon." *Endangered Species Research* 44: 291–325.

Roman, Joe, Irit Altman, Meagan M. Dunphy-Daly, Caitlin Campbell, Michael Jasny, and Andrew J. Read. 2013. "The Marine Mammal Protection Act at 40: Status, Recovery, and Future of US Marine Mammals." *Annals of the New York Academy of Sciences* 1286 (1): 29–49.

Southall, Brandon L., Douglas P. Nowacek, Ann E. Bowles, Valeria Senigaglia, Lars Bejder, and Peter L. Tyack. 2021. "Marine Mammal Noise Exposure Criteria: Assessing the Severity of Marine Mammal Behavioral Responses to Human Noise." *Aquatic Mammals* 47 (5): 421–464.

Marine Mammal Stranding Networks

Marine mammal stranding networks are coordinated groups of trained responders, including scientists, veterinarians, nonprofit organizations, government agencies, and community volunteers, who assist marine mammals who are stranded due to injury, entanglement, illness, or separation from their mothers. A stranding occurs when animals such as whales, dolphins, or sea lions become trapped on beaches or shallow waters and cannot return to the ocean without assistance. Organized responses to such events began to develop in the late twentieth century as governments and marine scientists recognized the scientific and conservation value of documenting stranded animals. In the US, these efforts were formalized through the Marine Mammal Protection Act (MMPA) and expanded in the 1990s through NOAA's Marine Mammal Health and Stranding Response Program, which coordinates a nationwide network of more than one hundred partner organizations (→ Marine Mammal Protection Act) (Gulland et al. 2025). Today, marine mammal stranding networks operate in many coastal regions around the world, including North America, Europe, New Zealand, and parts of Asia, serving as an important bridge between marine science, wildlife management, and community engagement (see Petitguyot et al. 2025; Wund et al. 2023).

A well-known example is the US National Marine Mammal Stranding Response Network that collaborates with universities, aquariums, wildlife rehabilitation centers, and local agencies to respond to stranded animals. When a stranding is reported, trained responders assess the animal's condition, collect biological samples, and determine whether the animal can be rehabilitated and released. If an animal has died, researchers may conduct necropsies to identify causes of mortality such as disease, pollution exposure, vessel strikes, or entanglement in fishing gear. Similar systems exist elsewhere, such as New Zealand's Department of Conservation stranding response teams that coordinate volunteers and marine mammal medics to refloat stranded animals during mass strandings (Palmer et al. 2022). Beyond emergency response, these networks maintain long-term databases of strandings that help scientists track patterns in marine mammal health, mortality, and environmental change.

Marine mammal stranding networks also play a significant role in sustainability because they generate critical information about the health of marine ecosystems (→ Sustainability). Marine mammals are widely regarded as "sentinel species," which means that changes in their health or mortality can reveal broader ecological problems such as climate change, pollution, harmful algal blooms, or shifts in prey availability (→ Climate Change) (→ Pollution) (→ Sentinels). Data collected from stranded animals, including tissue samples, disease testing, and population records, allow researchers to monitor environmental stressors and detect emerging threats to ocean biodiversity (→ Biodiversity). In this way, stranding networks function as both conservation tools and early warning systems for ocean sustainability (Lennon et al. 2025). By combining scientific research, wildlife rescue, and community participation, these networks help societies better understand human impacts on marine ecosystems and support more informed approaches to protecting marine biodiversity and ocean health.

References

Gulland, Frances M.D., Regina Asmutis-Silvia, Jeff Boehm, et al. 2025. "Marine Mammal Stranding Networks in the 21st Century: Whence and Whither?" *Marine Mammal Science* 41 (3): e70016.

Lennon, Rachel L., Rosie S. Williams, Kathryn J. Allan, et al. 2025. "An Approach to Using Stranding Data to Monitor Cetacean Population Trends and Guide Conservation Strategies." *Scientific Reports* 15 (1): 28417.

Palmer, Emily, Alana Alexander, Libby Liggins, et al. 2022. "A Piece of the Puzzle: Analyses of Recent Strandings and Historical Records Reveal New Genetic and Ecological Insights on New Zealand Sperm Whales." *Marine Ecology Progress Series* 690: 201–217.

Petitguyot, Marie A.C., Andrea Fariñas-Bermejo, Andrew Brownlow, et al. 2025. "European Stranding Networks as a Tool for Monitoring Marine Mammal Populations (Part I): Towards Optimising the Functioning of Networks." *ICES Journal of Marine Science* 82 (11): fsaf194.

Wund, Sarah, Eleonore Méheust, Cécile Dars, et al. 2023. "Strengthening the Health Surveillance of Marine Mammals in the Waters of Metropolitan France by Monitoring Strandings." *Frontiers in Marine Science* 10: 1116819.

Mesopredator Release

The removal of top/apex predators often creates a top-down control mesopredator release process in which medium-sized predators increase and play greater roles in the ecosystem (→ Trophic Cascades). A mesopredator is any mid-ranking predator in a food web, and a mesopredator in one ecosystem may be a top predator in another (Prugh et al. 2009). For example, feral cats function as mesopredators in some ecosystems, but in many island environments they are apex predators (→ Invasive Animal Species). Substantial ecological change from mesopredator release is most likely when the apex predator and mesopredator differ in body size and when the mesopredator is primarily a hunter (van Schaik et al. 2025). While mesopredator release is an important phenomenon in predator communities, it is only one of multiple potential responses by lower-ranked predators to changes in higher-ranked predator populations (→ Predation Paradox) (→ Trophic Cascades) (→ Ecological Traps) (van Schaik et al. 2025).

Mesopredator release is often characteristic of a fundamental ecosystem imbalance, such as the control of raccoons in Florida to protect sea turtle eggs that resulted in increased predation on the eggs by the ghost crab who was previously controlled by raccoons (Prugh et al. 2009). In a classic study of the consequences of an ecosystem imbalance in the human-made island environment of Lago Guri in Venezuela, Terborgh et al. (2001) document the impact of the loss of jaguars, pumas, and eagles due to the fragmentation of the land by the flooding of a hydroelectric dam. The absence of those apex predators increased the population of herbivore mesopredators (howler monkeys, rodents, iguanas, and leaf-cutter ants) who preyed on the island's birds, lizards, anurans, and spiders. This devastated the island vegetation and stripped out the forest understory. Mammalian mesocarnivores, in particular, are important drivers of ecosystem function, structure, and dynamics in three situations: 1) when larger carnivores are absent, 2) on island ecosystems or localities with simple community composition, and 3) where they are nonnative introductions (→ Trophic Cascades) (→ Island Biogeography) (→ Invasive Animal Species) (Roemer et al. 2009).

Mesopredator release can be advantageous in some ecosystems. For example, in the coral reef environment, the presence of sharks significantly affects the abundance and behavior of small benthic rays who play a critical role in the sustainability of the benthic ecosystem by oxygenating sediments through bioturbation (→ Animal Forests) (Sherman et al. 2020). However, rays also prey on invertebrates such as clams, snails, and scallops. The human overharvesting of sharks in the coastal northwest Atlantic ecosystem led to the explosion of the cownose ray that, along with environmental events and fishing, reduced bay scallop populations such that North Carolina's century-old scallop fishery was forced to close in 2004 (Myers et al. 2007) and is currently closed indefinitely to allow the population to increase.

References

Myers, Ransom A., Julia K. Baum, Travis D. Shepherd, Sean P. Powers, and Charles H. Peterson. 2007. "Cascading Effects of the Loss of Apex Predatory Sharks from a Coastal Ocean." *Science* 315: 1846–1850.

Prugh, Laura R., Chantal J. Stoner, Clinton W. Epps, et al. 2009. "The Rise of the Mesopredator." *BioScience* 59 (9): 779–791.

Roemer, Gary W., Matthew E. Gompper, and Blaire Van Valkenburgh. 2009. "The Ecological Role of the Mammalian Mesocarnivore." *BioScience* 59 (2): 165–173.

Sherman, C. Samantha, Michelle R. Heupel, Stephen K. Moore, Andrew Chin, and Colin A. Simpfendorfer. 2020. "When Sharks are Away, Rays Will Play: Effects of Top Predator Removal in Coral Reef Ecosystems." *Maine Ecology Progress Series* 641: 145–157.

Terborgh, John, Lawrence Lopez, Percy Nuñez V., et al. 2001. "Ecological Meltdown in Predator-Free Forest Fragments." *Science* 294: 1923–1926.

van Schaik, Thomas, Marijke van Kuijk, and Elisabeth H.M. Sterck. 2025. "Understanding Mesopredator Responses to Changes in Apex Predator Populations in Europe: Implications for the Mesopredator Release Hypothesis." *Mammal Review* 55: e12357.

Metabolic Rift Theory

Metabolic rift theory draws on Karl Marx's analysis of agriculture under capitalism (Foster 1999; Foster et al. 2010). Under capitalism, agriculture is driven by the desire for profit and the need to compete in a market, so depletion of soil health and other adverse environmental consequences tend to be ignored. Since the consumption of agricultural products takes place largely in urban settings, there are rifts between where resources are extracted, where the resulting products are consumed, and where ultimate waste products are disposed (→ Telecoupling). Rift theory has a similar logic to the idea of "externality" in conventional economics. Many sustainability consequences of economic activity are "external" to the market and are therefore not reflected in prices and not given any value in the market prices that drive decision making (Dietz 2023). Metabolic rift theory emphasizes the politics that shape these dynamics, arguing that capitalism produces a fundamental rift that disrupts the sustainability of ecosystem metabolism (→ Sustainability).

The approach has been applied to analyze animal agriculture, aquaculture, and fisheries (→ Livestock Production) (→ Aquaculture) (Gunderson 2011; Longo et al. 2015; Gunderson 2013; Foster and Clark 2018). Altmire and York have offered an analysis of the "anthophilic rift" using the term anthophilic because *Anthophila* is the scientific name for the clade of bees, with the prefix antho- meaning "flower" and the suffix -phila meaning "loving" (Altmire and York 2026, 2). They argue that the spread of settler colonialism and capitalism throughout much of the world led to the spread of the western honeybee (*Apis mellifera*) and the displacement of many native species (→ Colonialism) (→ Native Species). That spread in turn has facilitated the pollinator crisis — a dearth of pollinators in many ecosystems that threatens biodiversity and ecosystem dynamics (→ Biodiversity).

References

Altmire, Gabriella R., and Richard York. 2026. "The Anthophilic Rift: Advancing a Sociology of Biodiversity Loss through the Pollination Crisis." *Environmental Sociology*, February 5, 1–14. https://doi.org/10.1080/23251042.2026.2626279.

Dietz, Thomas. 2023. *Decisions for Sustainability: Facts and Values*. Cambridge University Press.

Foster, John Bellamy. 1999. "Marx's Theory of Metabolic Rift: Classical Foundations for Environmental Sociology." *American Journal of Sociology* 105 (2): 366–405.

Foster, John Bellamy, and Brett Clark. 2018. "The Robbery of Nature: Capitalism and the Metabolic Rift." *Monthly Review* 70 (3): 1–20.

Foster, John Bellamy, Brett Clark, and Richard York. 2010. *The Ecological Rift: Capitalism's War on the Earth*. Monthly Review Press.

Gunderson, Ryan. 2011. "The Metabolic Rifts of Livestock Agribusiness." *Organization & Environment* 24 (4): 404–422.

Gunderson, Ryan. 2013. "From Cattle to Capital: Exchange Value, Animal Commodification, and Barbarism." *Critical Sociology* 39 (2): 259–275.

Longo, Stefano B., Rebecca Clausen, and Brett Clark. 2015. *The Tragedy of the Commodity: Oceans, Fisheries, and Aquaculture*. Rutgers University Press.

Multispecies Futures

Multispecies futures is an interdisciplinary idea that considers how people can imagine and create thriving environments shared in coexistence with other species, including fungi, microorganisms, animals, and plants. It challenges human-centered perspectives (→ Anthropocentrism), highlights the roles all species play in shaping a "more-than-human-world," and recognizes the interconnectedness of all life for their sustainable futures (→ Biocentrism). In paying attention to different species, multiple connections and understandings are revealed through a range of ethnographic and ethological approaches, including engagement and collaboration with scientists, farmers, hunters, Indigenous people, activists, and artists (→ Citizen Science) (→ Indigenous and Decolonial Perspectives) (van Dooren et al. 2016). The multispecies approach aims to build inclusive environments where humans, non-human species, and ecosystems can all thrive together, promoting care and justice for every living entity (→ Environmental Justice) (→ Ecocentrism).

A notable example in the study from a multispecies futures approach is the collaborative hunting interactions of the day octopus with fish species such as the coral trout and the goatfish. Sampaio and colleagues (2025) assert that, while the octopus has traditionally been considered the nucleus of the interactions with the fish who merely follow along to opportunistically capture missed prey, recent studies have challenged that model. Rather than simply trailing behind, the fish take an active role in the hunting partnership by locating prey and visually signaling the location to the octopus who then uses her arms to envelop elements of the substrate so that she can ensnare or flush the prey from the hiding spot, thus enhancing the capture of prey for both parties (Sampaio et al. 2025). This collaboration illustrates the octopus' ability to interpret both subtle behavioral cues (locating prey or passive communication) and signals (active communication) from the fish (→ Cognition) (→ Learning for Survival) (→ Cultural Transmission) (Sampaio et al. 2025). This example highlights the advanced cognitive abilities in interspecies coordination and communication. When octopuses and fish team up to hunt, their coordinated strategies help them find and catch prey more efficiently. By cooperating, they save time and energy in both searching for food and capturing it, leading to greater foraging success for both species (Sampaio et al. 2025).

Cattle herding is also a multispecies activity where horse, cattle, and human must learn to communicate and develop a shared understanding of what it means to move conjointly (Pettitt and Brandt-Off 2022). Each species is socialized (or "zoocialized") to learn how to participate in a shared community of communication and a sense of the "timing" and "feel" of the other species to enable movement together, thus constructing a "vibrant multispecies community of communication" (→ Learning for Survival) (Pettit and Brandt-Off 2022, 178). Specific processes of zoocialization are always intersectional in that they depend on place, time, and culture and engage species, gender, ethnicity, and class (→ Intersectional Environmentalism), and knowing more about how humans and animals construct a shared understanding has implications for animal welfare particularly for those who will become food (→ Welfare) (Pettit and Brandt-Off 2022).

In proposing the concept of a "multispecies sustainability," Rupprecht et al. (2020) argue that since living beings depend on each other for their well-being, true sustainability can only be achieved if the interdependent needs of all species of current and future generations are met. The concept of multispecies futures encourages us to rethink our connections and ethics to support coexistence, restore damaged ecosystems, and build fairer, more robust societies that welcome diverse life forms (→ Ecocentrism).

References

Petitt, Andrea, and Keri Brandt-Off. 2022. "Zoocialization: Learning Together, Becoming Together in a Multispecies Triad." *Society & Animals* 32 (2): 178–195.

Rupprecht, Christoph D.D., Joost Vervoort, Chris Berthelsen, et al. 2020. "Multispecies Sustainability." *Global Sustainability* 3 (e34): 1–12.

Sampaio, Eduardo, Alexandra K. Schnell, and Piero Amodio. 2025. "Cognition in Multi-Species Sociality." *Current Biology* 35: R125–R138.

van Dooren, Thom, Eben Kirksey, and Ursula Münster. 2016. "Multispecies Studies: Cultivating Arts of Attentiveness." *Environmental Humanities* 8 (1): 1–23.

N

Native Species

Native species are organisms who are naturally found in a particular geographic region without human introduction. Their histories, ecological roles, and interactions have evolved over long periods of time, producing relationships that help stabilize ecosystems, shape food webs, and support nutrient cycles. Native species have often adapted to climate conditions, soils, predators, and symbiotic partners. Contemporary ecological scholarship increasingly questions the use of "native" as a categorical descriptor, noting that such classifications depend on contested historical baselines, dynamic ecological conditions, and the pervasive influence of human activity (→ Applied Zooarchaeology). Moreover, the native-non-native distinction often carries implicit normative judgments, positioning native species as worthy of protection while framing non-native species as legitimate targets of removal (Lynch et al. 2025). Biodiversity conservation strategies require a more nuanced approach than prioritizing the rarest, most vulnerable, charismatic, or native species (→ Biodiversity) (→ Conservation Biology) (→ Foundation Species). Growing research suggests that preserving ecological and evolutionary processes is central to success (Tobias et al. 2025). Understanding native species is an exercise in understanding both evolutionary history and current ecological contexts (→ Applied Zooarchaeology).

The idea of understanding native species in terms of ecosystem services and evolutionary processes becomes clearer when applied to non-human animals where range is a key feature in their ability to thrive (→ Ecosystem Services). For example, keystone pollinators such as the rusty patched bumble bee (*Bombus affinis*) in North America support the reproduction of numerous native plants, many of which provide habitat and food for other wild animals. When non-native or invasive species, such as the European honeybee, are introduced and compete with native pollinators, the ripple effects can undermine entire ecosystems (→ Invasive Animal Species) (→ Trophic Cascades). Recent studies show that declines in native pollinators reduce fruiting success for wild plants and disrupt food sources for birds, mammals, and other insects who depend on those resources (Martén-Rodríguez et al. 2025). This example illustrates how native

DOI: 10.4324/9781003528142-14

species contribute to ecosystem stability and how non-native species can reshape ecological communities.

Native species are central to sustainability because they can help maintain biodiversity, support resilient ecosystems, and contribute to climate change adaptation (Malhi et al. 2022). However, climate change is altering temperature patterns, precipitation cycles, and habitat availability, causing many species to shift their ranges in search of more suitable conditions (→ Climate Change). As a result, distinctions between native and non-native species are becoming more complicated, as some species are moving into new regions naturally, while others may become invasive under rapidly changing climate conditions (Essl et al. 2020). Sustainable environmental planning must navigate the dual challenge of protecting native species and managing non-native species in ways that support ecological integrity. This includes habitat restoration, climate-informed conservation strategies, and policies that anticipate shifting species distributions while safeguarding the ecological roles of native species.

References

Essl, Franz, Bernd Lenzner, Sven Bacher, et al. 2020. "Drivers of Future Alien Species Impacts: An Expert-Based Assessment." *Global Change Biology* 26 (9): 4880–4893.

Lynch, Kate E., Benjamin L. Allen, Oded Berger-Tal, et al. 2025. "Explicit Value Trade-Offs in Conservation: Integrating Animal Welfare." *Trends in Ecology & Evolution* 40 (6): 593–600.

Malhi, Yadvinder, Tonya Lander, Elizabeth le Roux, et al. 2022. "The Role of Large Wild Animals in Climate Change Mitigation and Adaptation." *Current Biology* 32 (4): R181–196.

Martén-Rodríguez, Silvana, Edson Jacob Cristobal-Pérez, Martín Hesajim de Santiago-Hernández, et al. 2025. "Untangling the Complexity of Climate Change Effects on Plant Reproductive Traits and Pollinators: A Systematic Global Synthesis." *Global Change Biology* 31 (2): e70081.

Tobias, Joseph A., James M. Bullock, Lynn V. Dicks, Brenna R. Forester, and Orly Razgour. 2025. "Biodiversity Conservation Requires Integration of Species-Centric and Process-Based Strategies." *Proceedings of the National Academy of Sciences* 122 (31): e2410936122.

Naturalist Programming

Naturalist programs are educational activities and public engagement efforts designed to help people learn about the natural world through guided observation, interpretation, and hands-on interaction with lands, plants, and animals. These programs are typically led by trained naturalists. Naturalists are educators, researchers, or community volunteers who introduce participants to plants and animals through guided activities and interpretation. Naturalist programming often takes place in zoos, aquariums, nature centers, and coastal or forested environments where participants can have direct encounters with nature. Rather than relying solely on classroom instruction, these programs emphasize experiential learning, encouraging people to observe species, ask questions, and develop a deeper understanding of ecological relationships. This approach draws on the long tradition of natural history education, which has historically played an important role in cultivating ecological literacy and curiosity about the living world (Tewksbury et al. 2014; Pattison and Dierking 2013).

Naturalist programming appears in many forms. National and state parks, for example, often offer interpretive programs led by park naturalists that include guided walks, wildlife talks, and habitat exploration activities. Zoos and aquariums frequently host naturalist-style encounters where visitors learn about animal behavior, conservation challenges, and ecosystem dynamics. Community-based programs also play an important role. Master naturalist programs train volunteers in ecology, conservation, and science communication so they can support habitat restoration projects, lead nature walks, and assist with environmental education (→ Ecology). In coastal communities, beach naturalist programs provide a particularly visible example of this approach. Volunteers or staff naturalists interact with visitors during low tides, helping them identify and better appreciate coastal ecosystems. These programs often combine informal conversations, species identification, and storytelling to create memorable learning experiences that connect visitors with marine biodiversity (Ballard et al. 2017; Ballantyne et al. 2007). Some organizations are going a step further to not only educate and spark curiosity among visitors, but to also enhance empathic concern for lesser known and less charismatic creatures (→ Empathy). As an example, the Seattle Aquarium trains its volunteer naturalists on how to talk about coastal shoreline creatures to not only impart information, but to also encourage perspective taking and empathic concern (Fennell 2024).

Naturalist programming is closely connected to sustainability because it is designed to help people develop knowledge, values, and behaviors that support environmental stewardship (→ Values). Research on environmental education consistently shows that direct experiences with nature can strengthen ecological understanding and encourage pro-environmental attitudes (DeVille et al. 2021). Programs that involve observation, dialogue, and participation, such as guided nature walks or citizen science activities, can deepen participants' sense of connection to ecosystems and motivate conservation action (→ Citizen Science) (Merenlender et al. 2016). Naturalist programming also contributes to sustainability by translating scientific knowledge into accessible forms for diverse audiences, making complex ecological issues understandable in everyday contexts. By helping people notice wildlife, understand ecosystems, and reflect on human relationships with nature, naturalist programs build the social foundations necessary for conservation and sustainable environmental management (Ardoin et al. 2020; Ballantyne et al. 2007; Tewksbury et al. 2014).

References

Ardoin, Nicole M., Alison W. Bowers, and Estelle Gaillard. 2020. "Environmental Education Outcomes for Conservation: A Systematic Review." *Biological Conservation* 241: 108224.

Ballantyne, Roy, Jan Packer, Karen Hughes, and Lynn Dierking. 2007. "Conservation Learning in Wildlife Tourism Settings: Lessons from Research in Zoos and Aquariums." *Environmental Education Research* 13 (3): 367–383.

Ballard, Heidi L., Lucy D. Robinson, Alison N. Young, et al. 2017. "Contributions to Conservation Outcomes by Natural History Museum-Led Citizen Science: Examining Evidence and Next Steps." *Biological Conservation* 208: 87–97.

DeVille, Nicole V., Linda Powers Tomasso, Olivia P. Stoddard, et al. 2021. "Time Spent in Nature Is Associated with Increased Pro-Environmental Attitudes and Behaviors." *International Journal of Environmental Research and Public Health* 18 (14): 7498.

Fennell, David A. 2024. "Empathy in Animal-Based Tourism Contrasting Constructed Care and Care Ethics at a Captive Wildlife Venue." *Tourism Recreation Research* 49 (6): 1222–1237.

Merenlender, Adina M., Alycia W. Crall, Sabrina Drill, Michelle Prysby, and Heidi Ballard. 2016. "Evaluating Environmental Education, Citizen Science, and Stewardship through Naturalist Programs." *Conservation Biology* 30 (6): 1255–1265.

Pattison, Scott A., and Lynn D. Dierking. 2013. "Staff-Mediated Learning in Museums: A Social Interaction Perspective." *Visitor Studies* 16 (2): 117–143.

Tewksbury, Joshua J., John G.T. Anderson, Jonathan D. Bakker, et al. 2014. "Natural History's Place in Science and Society." *BioScience* 64 (4): 300–310.

O

One Health

One Health is defined by the World Health Organization (WHO) as a transdisciplinary approach to sustainably balance and optimize the closely linked and interdependent health of people, domestic and wild animals, plants, and ecosystems (→ Domestic Animals) (→ Wilderness Animals) (→ Ecosystems). One Health focuses on the prevention and mitigation of a range of health issues, including zoonotic diseases, vector-borne diseases (Lyme, West Nile virus, malaria), food safety caused by contamination of food during production, and environmental health threats such as pollution and climate change (→ Zoonosis) (→ Livestock Production) (→ Pollution) (→ Climate Change). One Health's transdisciplinary approach draws on the participation of multiple disciplines (veterinary health, human health, environmental science, ecology), stakeholders, and communities; follows the Precautionary Principle that calls for precaution in the face of uncertainty; and recognizes the interconnectedness of systems (environmental, sociocultural, economic) (→ Citizen Science) (→ Coupled Human and Natural Systems) (Prata et al. 2022).

There are five key underlying principles of One Health (Adisasmito et al. 2022, 3): 1) equity between sectors and disciplines; 2) sociopolitical and multicultural parity (the doctrine that all people are equal and deserve equal rights and opportunities) and inclusion and engagement of communities and marginalized voices (→ Indigenous and Decolonial Perspectives); 3) socioecological equilibrium that seeks a harmonious balance between human-animal-environment interactions and acknowledges the importance of biodiversity, access to sufficient natural space and resources, and the intrinsic value of all life in the ecosystem (→ Biocentrism); 4) stewardship and the responsibility of humans to change behavior and adopt sustainable solutions that recognize the importance of animal welfare and the integrity of the whole ecosystem, thus securing the well-being of current and future generations (→ Sustainable Development Goals); and 5) transdisciplinarity and multisectoral collaboration that includes all relevant disciplines, both modern and traditional forms of knowledge and a broad representative array of perspectives (→ Citizen Science).

Pastoral communities have long engaged with the concept of One Health in their veneration and preservation of animals and the environment and the

DOI: 10.4324/9781003528142-15

rational use of natural resources (→ Indigenous and Decolonial Perspectives) (Prata et al. 2022). With the onset of two pandemics in the twenty-first century (the severe acute respiratory syndrome in 2002 and the coronavirus disease in 2020), the interconnection between the health of humans, animals, and their shared ecosystem highlighted the need for a One Health approach to address global health.

Concepts similar to One Health have emerged to address the impacts of ecological degradation on global health, specifically EcoHealth and Planetary Health. In a study of the foci of the three approaches, Talukder et al. (2024) conclude that EcoHealth focuses on the ecological, social, and economic aspects of health (→ Conservation Finance for Animals and Ecosystems), Planetary Health explores the health of human civilization and its interdependent relationship with the planet (→ Anthropocentrism), and One Health recognizes the interconnectedness of human, animal, and environmental health (→ Ecocentrism). The three approaches align in their acknowledgment of the "significance of ecosystems, climate change, sustainability, health-ecosystem interdependence, and the role of ecological and social determinants in health" (Talukder et al. 2024, 73).

References

Adisasmito, Wiku B., Salama Almuhairi, Casey Barton Behravesh, et al. 2022. "One Health: A New Definition for a Sustainable and Healthy Future." *PLoS Pathogens* 18 (6): e1010537. https://doi.org/10.1371/journal.ppat.1010537.

Prata, Joana C., Ana Isabel Ribeiro, and Teresa Rocha-Santos. 2022. "An Introduction to the Concept of One Health." In *One Health: Integrated Approach to 21st Century Challenges to Health*, edited by Joana C. Prata, Ana Isabel Ribeiro and Teresa Roha-Santos, 1–31. Elsevier.

Talukder, Byomkesh, Nilanjana Ganguli, Eunice Choi, et al. 2024. "Exploring the Nexus: Comparing and Aligning Planetary Health, One Health, and EcoHealth." *Global Transitions* 6: 66–75. https://doi.org/10.1016/j.glt.2023.12.002.

P

Pets Evacuation and Transportation Standards Act (PETS)

The Pets Evacuation and Transportation Standards Act (2006) is a piece of US legislation that requires state and local emergency preparedness plans to account for the needs of pets and service animals in disasters (→ Disasters). The PETS Act includes domestic companion animals such as dogs, cats, birds, and certain small mammals, as well as service animals who assist people with disabilities. It excludes farmed animals (→ Livestock Production), research animals (→ Animal Research), and most wildlife (American Veterinary Medical Association 2022; Chadwin 2017). The legislation was enacted in 2006 after the widespread abandonment and loss of companion animals from Hurricane Katrina. Hurricane Katrina made landfall in Louisiana on August 29, 2005, as a Category 3 hurricane. It is estimated that somewhere between 100,000 and 250,000 pets were abandoned or left behind during the event. Much of this was because there were no pet-inclusive evacuation plans in place, shelters did not permit pets, people were ordered by officials to abandon pets, and many left pets behind thinking they would be able to return within a few days.

Reflecting on this event and others, scholars like Leslie Irvine (Irvine 2009; Irvine and Andre 2023) have shown how disasters can expose society's devaluation of animals, and how this devaluation leads to widespread social consequences for human and non-human lives. Specifically, using the case of Hurricane Katrina, Irvine (Irvine 2009) found that people often risk their own safety to save their companion animals (→ Domestic Animals) during disasters and in the aftermath of disasters, and that when policy negates the importance of non-human animal lives, humans suffer as well.

Ultimately, the PETS Act is more than a law. It is a symbol of sustainable ethics in practice. The relocation of shelter animals before hurricanes, or the inclusion of animals in wildfire evacuation planning shows a recognition of the interdependence of human and non-human animal communities. Preparing for the future, we must integrate animals into our visions of resilience. Sustainability is not just about preserving ecosystems but about building systems of care (→ Ecofeminism) that consider the needs and importance of all species (→ Ecocentrism). By integrating animals into policy frameworks like

DOI: 10.4324/9781003528142-16

the PETS Act, society begins to acknowledge that sustainability is not just about protecting resources, but about interspecies relationships. The PETS Act represents meaningful progress in recognizing human-animal relationships in disasters, yet its protections remain limited in scope.

References

American Veterinary Medical Association. 2022. "PETS Act (FAQ)." American Veterinarian Medical Association. https://www.avma.org/pets-act-faq.

Chadwin, Robin. 2017. "Evacuation of Pets during Disasters: A Public Health Intervention to Increase Resilience." *American Journal of Public Health* 107 (9): 1413–17.

Irvine, Leslie. 2009. *Filling the Ark: Animal Welfare in Disasters*. Temple University Press.

Irvine, Leslie, and Casara Andre. 2023. "Pet Loss in an Urban Firestorm: Grief and Hope after Colorado's Marshall Fire." *Animals* 13 (3): 416.

Pets Evacuation and Transportation Standards Act (PETS) 109–308 Pub. L. 2006. https://www.govinfo.gov/app/details/STATUTE-120/STATUTE-120-Pg1725.

Planetary Limits

The concept of planetary limits is based on the argument that the resource demands of a growing human population would outstrip available resources. The notion of planetary limits is usually traced to Thomas Malthus' *Essay on Population* (first published anonymously in 1798), but the claim that large human populations can create problems is even older (Dietz and Rosa 1994; Malthus 1803). The conviction that human impacts on the biosphere would lead to "limits to growth" became a prominent part of what would become the discussion of sustainability in 1972 (Meadows et al. 1972).

The initial notion of "limits" was eventually expanded into the concept of "planetary boundaries" (Rockström et al. 2009). Nine global processes were identified that were fundamental stresses on how the biosphere functions and on the ability of ecosystems to support humans and other species (→ Ecosystem Services) (→ Global Environmental Change). The boundaries are estimates of when human stress on the environment via these processes becomes so great as to risk potentially irreversible environmental change, thus threatening the well-being of humans and other species. These boundaries are usually displayed as a series of concentric circles with outward projections from the center for each stress. An inner circle represents a level of stress that poses little risk to the biosphere, while outer circles represent increasing stress and risk. The most recent assessment indicates that human activities were exceeding a safe operating space with regard to seven of the nine boundaries: ocean acidification, freshwater use, land use change, climate change, modification of biogeochemical flows, introduction of novel substances into the environment, and changes in the integrity of ecosystems (Sakschewski et al. 2025). Only stratospheric ozone depletion and atmospheric aerosol pollutants were deemed to be at low-risk levels.

Donut economics builds on the planetary boundaries as an outer limit to the amount of stress humans should place on the environment (Raworth 2017; Fanning and Raworth 2025), but it reformulates the circle into the shape of a donut. The outer edge of the donut is where human activities are creating dangerous stress on the environment, that is, they are passing planetary boundaries. The space inside the donut is where human actions are not sufficient to provide for the well-being of all humans. The argument is that humans should try to act in the donut while avoiding both the hole and going beyond the outer edge. That is, there should be enough economic activity and resulting stress on the environment to allow for well-being for all humans but not so much as to cross planetary boundaries. These are sometimes called the ecological ceiling and the social foundation.

Most formulations of donut economics are anthropocentric in that the inner circle is defined in terms of inadequate human well-being (→ Anthropocentrism). However, the planetary boundaries idea, while also human-focused, emphasizes major and irreversible changes in the biosphere. Such changes are considered to make sustainability unachievable, with adverse effects on all species on Earth.

References

Dietz, Thomas, and Eugene A. Rosa. 1994. "Rethinking the Environmental Impacts of Population, Affluence and Technology." *Human Ecology Review* 1: 277–300.

Fanning, Andrew L., and Kate Raworth. 2025. "Doughnut of Social and Planetary Boundaries Monitors a World out of Balance." *Nature* 646 (8083): 47–56. https://doi.org/10.1038/s41586-025-09385-1.

Malthus, Thomas R. 1803. *An Essay on the Principle of Population*. Cambridge University Press.

Meadows, Donella H., Dennis L. Meadows, Jørgen Randers, and William Behrens III. 1972. *The Limits to Growth*. Universe Books.

Raworth, Kate. 2017. *Doughnut Economics: Seven Ways to Think like a 21st Century Economist*. Chelsea Green Publishing.

Rockström, Johan, Will Steffen, Kevin Noone, et al. 2009. "A Safe Operating Space for Humanity." *Nature* 461 (7263): 472–475. https://doi.org/10.1038/461472a.

Sakschewski, Boris, Levke Caesar, Lauren Andersen, et al. 2025. *Planetary Health Check 2025: A Scientific Assessment of the State of the Planet*. Potsdam Institute for Climate Impact Research (PIK). https://doi.org/10.48485/PIK.2025.017.

Pollution

Pollution is often defined as substances entering ecosystems as a result of human action. Some of these substances are toxic in the sense that they have adverse physiological effects on animal or plant health. Natural processes can expose animals to toxic substances, such as sulfur dioxide gas from volcanic eruptions, cadmium, lead, and other heavy metals leaching from mineral deposits, or fine soot from forest fires. However, with the exception of volcanic activity, human action often exacerbates these natural processes. Other substances that are considered pollution are not toxic but are released in quantities that overwhelm ecosystem processes leading to massive changes, including adverse effects on animals. For example, flows of animal waste in moderate amounts are part of normal processes in many ecosystems. However, when intensive animal agriculture greatly increases the volume of waste it overwhelms the ability of the ecosystem to process it, leading to depletion of oxygen in lakes and rivers and killing animal life (→ Livestock Production) (Blaszczak et al. 2023). In some lakes and rivers, this has led to large "dead zones" that persist even when the influx of pollution is reduced. Many of these processes are exacerbated by climate change (→ Climate Change).

Throughout human history, human actions have generated pollution. For example, traces of the large amounts of lead released by the Roman Empire can be detected in Greenland ice cores and lake sediments throughout Europe (McConnell et al. 2025). Pollution was greatly amplified by the Industrial Revolution, colonialism, and the increased use of fertilizer and pesticides in agriculture (→ Colonialism). Advances in chemistry led to the development of new substances, some of which, like pesticides, were designed to be toxic to targeted species. Intentionally toxic substances often have unintended impacts beyond their target species. DDT was highly effective at killing targeted insect species and has relatively low toxicity to humans, but it interferes with the chemistry of egg production in birds with devastating results, including the near extinction of bald eagles and several other raptors (→ Extinction) (Dunlap 2014). Even non-toxic pollution can have devastating effects. Chlorofluorocarbons have no direct impact on human and animal health, but they degrade the layer of ozone in the upper atmosphere that shields the Earth's surface from ultraviolet radiation and thus have a devastating indirect effect on ecosystems. Carbon dioxide and methane are naturally occurring compounds with low toxicity, but human actions have released them in such massive amounts that they are altering the climate (→ Climate Change).

Many toxins have two features that make them particularly disruptive. First, they are persistent – unlike materials produced by biological processes, ecosystems have no evolved mechanisms to break them down. Some of them are referred to as POPs (Persistent Organic Pollutants) to emphasize that they are long-lived. PFASs (per- and polyfluoroalkyl substances) are so resistant to breaking down in the environment that they are called "forever chemicals," and it is argued that the risk from them constitute a planetary boundary (→ Planetary Boundaries) (Weber and Girones 2026). Second, many toxins bioaccumulate (Tansel 2026). Plants may absorb them at low concentrations, but when herbivores eat the plants, the concentration increases. When first-level predators eat the herbivores, the concentration increases again, and so on up the food chain, to the top predators, including humans. Thus, the animals most heavily exposed are top predators (→ Keystone Species). Such bioaccumulation has created serious problems for raptors, as noted above, for some marine mammals, and for ocean predators such as tuna and swordfish. In turn, as humans tend to eat animals near the top of the food chain, these concentrations can create health risks for them as well. There is also growing evidence that PFAS chemicals and microplastics (microscopic pieces of plastic) not only present risks on their own but also act as carriers for toxic chemicals. Overall, the effects of pollutants are so serious that they are considered part of global environmental change (→ Global Environmental Change).

Pollution can also refer to human-caused increases in noise and lighting that interfere with animal adaptation and communication. Artificial light at night (ALAN) has many disruptive effects, especially on invertebrates (Machnowska and Skwarlo-Sonta 2026). ALAN throws off circadian rhythms: it interferes with pollinators by shifting the timing of plant flowering, changes predator-prey dynamics, and can disorient migratory birds, insects, and marine turtles. These processes harm individual animals but also disrupt ecosystems (→ Ecosystems). Increases in human noise also have broad impacts on many non-human animal species. Birds use acoustic cues to guide much of their behavior so high levels of noise have widespread disruptive effects on many species (Madden et al. 2026). Many marine species also make extensive use of sound, and human noise in the oceans is proving highly disruptive to them (Harding and Cousins 2022).

Concern with how pesticides, particularly DDT, were impacting animals motivated Rachel Carson to write *Silent Spring* (Carson 1962; Lear 1998). Carson articulated the vision for what became the modern environmental movement, linking the traditional conservation movement's concern with wildlife to the public health movement's concern with pollution. While some pollution threats to animals and sustainability have been addressed with policy, such as the banning in most countries of DDT and related pesticides, growing awareness of threats like microplastics and PFAS compounds as well as with noise and light pollution means that pollution remains a key issue for sustainability.

References

Blaszczak, Joanna R., Lauren E. Koenig, Francine H. Mejia, et al. 2023. "Extent, Patterns, and Drivers of Hypoxia in the World's Streams and Rivers." *Limnology and Oceanography Letters* 8 (3): 453–463. https://doi.org/10.1002/lol2.10297.

Carson, Rachel. 1962. *Silent Spring*. Houghton Mifflin Harcourt.

Dunlap, Thomas. 2014. *DDT: Scientists, Citizens, and Public Policy*. Princeton University Press.

Harding, Simon, and Neil Cousins. 2022. *Review of the Impacts of Anthropogenic Underwater Noise on Marine Biodiversity and Approaches to Manage and Mitigate Them*. CBD Technical Series No. 99, Secretariat of the Convention on Biological Diversity. www.cbd.int/doc/publications/cbd-ts-99-en.pdf.

Lear, Linda. 1998. *Rachel Carson: Witness for Nature*. Macmillan.
Machnowska, Agnieszka Anna, and Krystyna Skwarlo-Sonta. 2026. "Artificial Light at Night (ALAN), an Anthropogenic Pollutant: A Narrative Review." *Pollutants* 6 (1): 11. https://doi.org/10.3390/pollutants6010011.
Madden, Natalie L., Kirby L. Mills, Karen M. Alofs, Clinton D. Francis, and Neil H. Carter. 2026. "Trait-Mediated Effects of Anthropogenic Noise on Bird Behaviour and Fitness." *Proceedings of the Royal Society B: Biological Sciences* 293 (2064): 20252521. https://doi.org/10.1098/rspb.2025.2521.
McConnell, Joseph R., Nathan J. Chellman, Andreas Plach, et al. 2025. "Pan-European Atmospheric Lead Pollution, Enhanced Blood Lead Levels, and Cognitive Decline from Roman-Era Mining and Smelting." *Proceedings of the National Academy of Sciences* 122 (3): e2419630121. https://doi.org/10.1073/pnas.2419630121.
Tansel, Berrin. 2026. "Trophic-Level Accumulation and Transfer of Legacy and Emerging Contaminants in Marine Biota: Meta-Analysis of Mercury, PCBs, Microplastics, PFAS, PAHs." *Marine Pollution Bulletin* 222 (January): 118666. https://doi.org/10.1016/j.marpolbul.2025.118666.
Weber, Roland, and Lautaro Girones. 2026. "PFAS and Planetary Boundaries: Global Challenges for Waste Management and Policy Solutions." *Environmental Science & Technology* 60 (5): 3707–3715. https://doi.org/10.1021/acs.est.5c12334.

Predation Paradox

Urbanization has fundamentally altered trophic dynamics for wildlife communities living in metropolitan areas creating a "predation paradox." A predation paradox occurs when predator populations increase but their predation on prey species declines because of the greater availability of anthropogenic foods (Fischer et al. 2012). The predation paradox is a variation in top-down effects; the increased bottom-up forcing and reduced top-down control is central to understanding trophic interactions in metropolitan environments (→ Trophic Cascades) (Fischer et al. 2012). The predation paradox distorts the usual food chain dynamics of an ecosystem, with the "top down" control of prey population by predators reduced and an increase in bottom-up forcing through anthropogenic food availability. For example, high food density and low predation risk explain the global pattern of very high urban bird populations (Shochat 2004). The combination of high food predictability and low predation risk in cities changes bird foraging behavior such that bird populations exceed the carrying capacity of the urban environment, taking a heavy toll on the body condition and life span of many birds (Shochat 2004).

For some predators such as hawks, eagles, kites, and for birds including gulls and corvids, anthropogenic food sources include meat-packing waste, roadkill, pet food, and foods fed deliberately to wild animals by humans (→ Road Ecology) (→ Wildlife Tourism) (→ Conflict with Terrestrial Animals). Their regular consumption of anthropogenic foods can create an ecological trap (→ Ecological Traps) (Morozov 2022). Thus, despite the high densities of predators, especially mesopredators, predation is higher in non-urbanized areas (→ Mesopredator Release). Physical changes in urban habitats can lead to shifts in targeted prey. Warmer city temperatures shift breeding patterns, artificial light at night enables twilight and nocturnal foraging by diurnal birds of prey (→ Pollution), and some structural features of buildings enable unusual hunting techniques (→ Animal-Inclusive Design) (Morozov 2022).

The predation paradox can also occur in benthic marine systems (→ Animal Forests). A natural experiment in two urbanized river habitats of the Chesapeake Bay in the US was conducted to determine the intensity of predation on urban reefs and whether oyster reef restoration enhances or diminishes predation (Ramírez-Joseph 2024). While control by top predators did not shape the ecological communities of restored or degraded urbanized oyster reefs, top-down control was weakened in both. Species richness was greater in the unrestored oyster reefs, but mud crab mesopredators were 338 percent more abundant in the restored habitat than in the degraded river habitat (Ramírez-Joseph 2024). The study concluded that restoration in the absence of top predators created habitat complexity that provided more refuge areas and prey for mesopredators, an unexpected outcome suggestive of an aquatic predation paradox (→ Ecological Restoration).

The predation paradox adversely affects sustainability by disrupting ecological balances, leading to population fluctuations, increasing stress on prey species, and potentially resulting in local extinctions (→ Extinction). This phenomenon illustrates how human interventions, such as habitat fragmentation and the removal of apex predators, can inadvertently generate instability in ecosystems (→ Habitat Loss) (→ Defaunation) (→ Keystone Species).

References

Fischer, Jason D., Sarah H. Cleeton, Timothy P. Lyons, and James R. Miller. 2012. "Urbanization and the Predation Paradox: The Role of Trophic Dynamics in Structuring Vertebrate Communities." *Bioscience* 62 (9): 809–818.

Morozov, N.S. 2022. "The Role of Predators in Shaping Urban Bird Populations. 4. The Urban Predation Paradox and Its Probable Causes." *Biology Bulletin* 49 (9): 1406–1432.

Ramírez-Joseph, Lucía. 2024. "Is the Predation- Paradox All Dry? Assessing the Strength of Top-Down Community Regulation on Urbanized, Temperate Oyster Reefs." PhD Dissertation, University of North Carolina at Greensboro. www.proquest.com/docview/3054375823?fromopenview=true&pq-origsite=gscholar&sourcetype=Dissertations%20&%20Theses.

Shochat, Eyal. 2004. "Credit or Debit? Resource Input Changes Population Dynamics of City-Slicker Birds." *OIKOS* 106 (3), 622–626.

Property

Property refers to socially and legally recognized rights for how individuals, groups, or organizations may access, use, control, and transfer resources (→ Rights). There are three recognized categories of property: real (land and structures), personal (movable objects and material goods), and intellectual (artistic works and inventions). Property law establishes the obligations and limitations associated with ownership.

Historically, livestock and companion animals have fallen into the personal property category, where property rights provide the legal authority to breed, confine, sell, or euthanize non-human animals within the limits of animal welfare law (→ Welfare) (→ Animal Law). As a comparison, children or other humans who require sustained and long-term care are not considered property but are individual rights-bearing persons. Of course, it has been a long historical struggle to abolish the practice of treating some humans as property, and scholars have considered the parallels to the status of non-human animals. Parents, guardians, and caretakers do not own these humans, but instead hold legal obligations to care for, protect, and make decisions that are in their best interests. Some scholars argue that the personal property status of non-human animals reduces sentient non-human animal beings to resources that further facilitates their exploitation in agriculture, entertainment, and research (→ Sentience) (→ Livestock Production) (→ Animal Research) (→ Welfare) (Francione 2008; 2012). Even when welfare standards are strengthened, the underlying legal structure still frames non-human animals based on human use value (→ Anthropocentrism).

The idea of property has long structured human-animal relationships. As animals were deemed to have economic value, they became incorporated in property law (→ Animal Law) (Favre 2017). Because of their quantifiable economic value, only domestic animals used to produce food or other products were considered by the legal system. Domestic companion animals were excluded because they had no perceived economic value (→ Domestic Animals). Today, most people don't own the animals they eat, and companion animals are often considered to be family members. Changes in how people view and relate to non-human animals are driving conversations about expanding animal rights and challenging traditional property categories. In an interesting legal development, companion animals are being included in protective orders issued by courts in abusive situations, an approach that defines them as more than mere property (Fitzgerald 2025).

Legal scholar David Favre proposes a fourth category of "living property," to address the unique status of non-human animals who are classified as personal property, similar to furniture and other inanimate objects, despite their sentience (Favre 2009). In Favre's model, the "bundle of rights" held by owners would be partially transferred to non-human animals so that animals themselves could retain the right to bodily integrity and basic well-being. At the core of this model is the acknowledgment of animals as living sentient beings rather than objects (→ Sentience). This idea is directly related to how some countries have granted limited personhood or rights to certain animals or ecosystems (→ Ecodemocracy) (→ Rights) (Chapron et al. 2019). In addition, some have argued that problems around managing common pool resources (including non-human animals) arise because no individual owns the commons or any part of it. But a body of scholarship demonstrates that many commons, such as fisheries, grazing land, or irrigation water, have been managed sustainably for decades or centuries by communities without any individual having property rights in them (→ Commons Dilemma).

The connection between property, non-human animals, and sustainability is clear when examining food systems. Treating animals as property enables intensive and confined animal production which leads to wild animal habitat loss, excessive resource consumption, increased greenhouse-gas emissions, and pollution (→ Livestock Production) (→ Habitat Loss) (→ Pollution). Given that property law prioritizes human ownership interests, environmental harms associated with industrial animal production are often shifted to local communities and ecosystems, which creates environmental justice issues (→ Environmental Justice). In general, property rights, with or without non-human animal recognition, can contribute to environmental inequality and undermine ecological resilience and sustainability (Robbins 2019). Discussions about sustainability should include conversations to challenge property rights structures, acknowledge Indigenous and decolonial perspectives, and encourage the expansion of stewardship, ecological limit recognitions, and valuing non-human animals beyond their utility (→ Sustainability) (→ Indigenous and Decolonial Perspectives).

References

Chapron, Guillaume, Yaffa Epstein, and José Vicente López-Bao. 2019. "A Rights Revolution for Nature." *Science* 363 (6434): 1392–1393.

Favre, David. 2009. "Living Property: A New Status for Animals within the Legal System." *Marquette Law Review* 93: 1021.

Favre, David. 2017. "Animals as Living Property." In *The Oxford Handbook of Animal Studies*, edited by Linda Kalof. Oxford University Press.

Fitzgerald, Amy. 2025. "Animals as Covictims and More-Than-Property: The Inclusion of Companion Animals in Protection Order Statutes at the State Level in the United States." *Violence & Victims* 40(1): 110–134. https://doi.org/10.1891/VV-2024-0004.

Francione, Gary. 2012. *Animals Property & the Law*. Temple University Press.

Francione, Gary Lawrence. 2008. *Animals as Persons: Essays on the Abolition of Animal Exploitation*. Columbia University Press.

Robbins, Paul. 2019. *Political Ecology: A Critical Introduction*. Wiley.

Q

Queer Ecology

Queer ecology is an interdisciplinary field that combines ideas from queer theory, feminist studies, environmental humanities, and posthumanism to challenge heteronormative and binary productions of nature (including views of non-human animals) by highlighting and interrogating the fluidity and diversity of human and non-human animal life. Queer ecology asks us to reconsider what is "normal" or "natural" in nature and society, and to imagine new, more inclusive ways of relating to the world around us (Seymour 2020). Interdisciplinary in nature, queer ecological research and writing can be thought of as a conversation between philosophy, sociology, psychology, animal studies, science and technology studies, and environmental studies. Leading queer ecological scholars including Catriona Mortimer-Sandilands and Bruce Erickson (2010) assert that the foundations of queer ecology can be traced back to Michel Foucault's (1978) and Judith Butler's (1990) work on performativity. The evolution of queer ecology can be broken down into two waves, both of which consider non-human animals as a central part of the field.

The first wave (1990 to 2010) focused on connecting and extending eco-feminist and queer theory to push against imaginary binaries used to describe the world, such as nature/culture, human/animal, male/female, natural/unnatural, good/bad, and to highlight diversity and divergence across species and ecosystems (→ Ecofeminism). Early work centering animals was heavily influenced by Donna Haraway's (2003) literary explorations of the interconnectedness of human and animal lives and her idea of "natureculture" that challenges the separation of nature and culture. Literature from this wave often highlighted the existence of same-sex animal relationships among species such as penguins and dolphins.

The second wave (2010s–present) moves beyond deconstructing binaries and applies intersectional justice to non-human animals, advocating for our understanding of how people think about and engage with nature and non-human animals to be intersectional. This shift in focus is heavily influenced by ecocentrism and environmental justice (→ Ecocentrism) (→ Environmental Justice). Central to this shift is the idea of queer political culture (QPC) that reflects

DOI: 10.4324/9781003528142-17

the recognition of a socialized solidified political and/or activist identity that exists within LGBTQ+ communities (→ Advocacy) (Bernstein and Reimann 2001). Historically, QPC is thought to function by consolidating power within the LGBTQ+ community to resist oppressive social norms and policies that attempt to restrict or remove foundational rights. However, research suggests that QPC fosters greater engagement in advocacy not only for LGBTQ+ related issues, but also for broader social justice causes, including concern for the well-being of non-human animals (Whitley and Bowers 2023). Thus, QPC is inherently intersectional and a key component of the evolution of queer ecology.

Contemporary queer and trans scholars in queer ecological studies focus on extending key concepts and promoting intersectional justice. Scholars like Cleo Wölfle Hazard (2022) seek to expand queer ecological frameworks to be more inclusive of gender variance by promoting queer and trans ecology that centers gender variation in theorizing about human-nature and human-animal interactions. Some of this scholarship has reinvigorated first-wave animal comparisons by drawing attention to species that can change sex, such as clownfish and certain amphibians, as well as to male seahorses who are unique in the animal kingdom for their role in gestation. Others, like Piepzna-Samarasinha (2022), a queer disability scholar, argue that environmental justice movements have often failed to acknowledge or incorporate disability. She draws on a wealth of work addressing disability and animality (→ Disability) (Jenkins et al. 2020). Likewise, queer disaster scholars like Leo Goldsmith and Michelle Bell (2022) work to extend queer ecology in our understanding of the impacts and reactions to disasters, recognizing that animals are an important part of this equation (→ Disasters). Queer ecology continues to deconstruct binaries by building intersectional frameworks across gender, race, disability, sexuality, and species (→ Intersectional Environmentalism). Increasingly, queer ecology is not just a critical lens to evaluate constructions of nature, but a praxis for fostering more inclusive ways of addressing environmental and animal-related issues. Queer ecology advances sustainability by challenging anthropocentric and heteronormative assumptions, fostering inclusive community-based models of environmentalism, and promoting practices that honor the diversity and complexity of nature. By rejecting a singular "nature," it calls for policies and solutions that integrate social and ecological justice.

References

Bernstein, Mary, and Renate Reimann, eds. 2001. *Queer Families, Queer Politics: Challenging Culture and the State*. Columbia University Press.

Butler, Judith. 1990. *Gender Trouble: Feminism and the Subversion of Identity*. Routledge

Foucault, Michel. 1978. *The History of Sexuality, Volume 1: An Introduction*. Pantheon Books.

Goldsmith, Leo, and Michelle L. Bell. 2022. "Queering Environmental Justice: Unequal Environmental Health Burden on the LGBTQ+ Community." *American Journal of Public Health* 112 (1): 79–87.

Haraway, Donna J. 2003. *The Companion Species Manifesto: Dogs, People, and Significant Otherness*. Prickly Paradigm Press.

Hazard, Cleo Wölfle. 2022. *Underflows: Queer Trans Ecologies and River Justice*. University of Washington Press.

Jenkins, Stephanie, Kelly Struthers Montford, and Chloë Taylor, eds. 2020. *Disability and Animality: Crip Perspectives in Critical Animal Studies*. Routledge.

Mortimer-Sandilands, Catriona, and Bruce Erickson, eds. 2010. *Queer Ecologies: Sex, Nature, Politics, Desire*. Indiana University Press.

Piepzna-Samarasinha, Leah L. 2022. *The Future Is Disabled: Prophecies, Love Notes and Mourning Songs*. Arsenal Pulp Press.

Seymour, Nicole. 2020. "Queer Ecologies and Queer Environmentalisms." In *The Cambridge Companion to Queer Studies*, edited by Siobhan B. Somerville, 108–122. Cambridge University Press.

Whitley, Cameron T., and Melanie M. Bowers. 2023. "Queering Climate Change: Exploring the Influence of LGBTQ+ Identity on Climate Change Belief and Risk Perceptions." *Sociological Inquiry* 93 (3): 413–439.

R

Religion

Religion can be understood as a complex and multifaceted system of shared beliefs, values, norms, practices, and symbols through which individuals and communities seek purpose, interpret existence, establish moral frameworks, and connect to what they perceive as sacred or transcendent. It links humanity to the spiritual, supernatural, or natural world, providing a framework for understanding the universe, giving meaning to life, and fostering social cohesion through shared rituals, ethical principles, and collective traditions. However, religious differences between groups can sometimes lead to sharp and even violent conflicts (Deitch 2022).

Much of the work on the relationship between religion, animals, and sustainability is situated within religious studies. This work considers how humans form ethical relationships with non-human animals based on religious values, or the use and meaning of non-human animals in sacred texts. In this way, religion provides a powerful framework for shaping human-animal relationships by offering moral and symbolic systems through which people interpret the place of animals in the world, with many religions articulating clear guidelines for the treatment of animals. These guidelines range from prohibitions against harm to the use of non-human animals in ritual sacrifices. Through creation stories, metaphors, and sacred texts, religions often assign non-human animals moral or spiritual significance, positioning them as spiritual companions, family members, and ancestors. These stories help shape attitudes toward animals, inform diets, and influence conservation and sustainability practices.

In Christianity, animals and sustainability have frequently been understood through a dominion-based framework, where humans have supreme authority over non-human animals and the environment (→ Anthropocentrism). Contemporary Christian environmental theology, influenced by both critical scholars like Lynn White, Jr. (1967) and theologians like Sallie McFague (1993), draws on historical figures like St. Francis of Assisi to argue that humans are called to be caretakers of the Earth and its creatures as part of a divine covenant rather than to exploit them (→ Values). This approach links

DOI: 10.4324/9781003528142-18

sustainability with ethical responsibility where protecting animals and their habitats becomes a sacred duty.

In Buddhism, animals and sustainability are viewed through the principle of ahimsa (no harm), which emphasizes compassion for all sentient beings (→ Sentience). In addition, samsara, the cycle of birth, death, and rebirth, places humans and non-human animals as interchangeable moral companions (→ Altruism). This worldview frames ecological responsibility as a spiritual practice and encourages mindfulness in consumption (Finnigan 2017).

In Hinduism, animals are understood as sacred beings intricately bound to the cosmic order (dharma) who occupy essential roles within broader moral and spiritual frameworks. Like Buddhism, non-violence is encouraged. This worldview situates ecological sustainability as a spiritual responsibility, framing the protection of biodiversity as integral to maintaining spiritual and environmental harmony (Dash et al. 2024).

In Islam, animals are viewed as sentient beings created by Allah with intrinsic value and purpose deserving of kindness and respect. The Qur'an emphasizes human responsibility as Khalifah (stewards) of the Earth, obligating believers to maintain balance (mīzān) and prevent harm to animals and the environment. This frames care for animals and sustainability as expressions of faith and accountability (Rahman 2017).

In Indigenous religions, although perspectives are very diverse, animals are often viewed as kin and/or spiritual beings who share reciprocal relationships with humans (→ Indigenous and Decolonial Perspectives). This worldview emphasizes respect, gratitude, and responsibility, encouraging sustainable living practices. Environmental stewardship is framed as a sacred duty through the recognition that all forms of life are interconnected (→ Biocentrism) (Weatherdon 2022). In some traditions, efforts are made to incorporate the perspective of non-human animals in decisions (→ Ecodemocracy).

Jainism places animals within the moral community as sentient beings deserving of nonviolence, requiring adherents to actively avoid harming even the smallest forms of life in everyday practices. This ethic of radical restraint extends to sustainability through principles of minimal consumption and careful resource use, linking ecological protection directly to spiritual responsibility.

Beyond assessing the place of animals in various religions, an evolving line of research is beginning to assess whether (and how) animals have religious experiences. This work is situated among comparative religion and cognitive ethology studies and essentially focuses on documented examples of animals mourning the death of conspecifics. For instance, elephants have been observed holding vigil over deceased herd members by standing near, touching, and revisiting the bones of family members in what appears to be a grief response (Goldenberg and Wittemyer 2020). Similarly, orcas and dolphins engage in "post-mortem attentive behavior" and will carry the bodies of deceased calves for days or even weeks (Reggente et al. 2018;

Bearzi et al. 2018). This work draws heavily on Marc Bekoff's (2000; 2024) research on how animals demonstrate emotional and possibly spiritual capacities, such as mourning and grief. These ideas expand sustainability beyond material conservation to include ethical consideration of animals as sentient participants in ecological and moral communities (→ Sentience) (→ Biocertrism). This recognition could foster more compassionate and holistic approaches to the conservation and management of non-human animals to achieve sustainability goals.

Religion plays a central role in shaping how people understand and relate to animals and the environment by offering moral and spiritual guidance for living responsibly in the global community. Across many traditions, teachings emphasize compassion, stewardship, and balance, framing the care of animals and sustainability as interconnected expressions of ethical and spiritual life.

References

Bearzi, Giovanni, Dan Kerem, Nathan B. Furey, Robert L. Pitman, Luke Rendell, and Randall R. Reeves. 2018. "Whale and Dolphin Behavioural Responses to Dead Conspecifics." *Zoology* 128: 1–15.

Bekoff, Marc. 2000. "Animal Emotions: Exploring Passionate Natures: Current Interdisciplinary Research Provides Compelling Evidence That Many Animals Experience Such Emotions as Joy, Fear, Love, Despair, and Grief — We Are Not Alone." *BioScience* 50 (10): 861–870.

Bekoff, Marc. 2024. *The Emotional Lives of Animals: A Leading Scientist Explores Animal Joy, Sorrow, and Empathy —And Why They Matter*. New World Library.

Dash, Bibhudatta, G. Lakshmi Narasimham, and Priyanka Jinde. 2024. "Animals in Hinduism: Exploring Communication Beyond the Human Realm in Sacred Texts and Practices." *Journal of Dharma Studies* 7 (3): 213–228.

Deitch, Mora. 2022. "Is Religion a Barrier to Peace? Religious Influence on Violent Intrastate Conflict Termination." *Terrorism and Political Violence* 34 (7): 1454–1470.

Finnigan, Bronwyn. 2017. "Buddhism and Animal Ethics." *Philosophy Compass* 12 (7): e12424.

Goldenberg, Shifra Z., and George Wittemyer. 2020. "Elephant Behavior toward the Dead: A Review and Insights from Field Observations." *Primates* 61 (1): 119–128.

McFague, Sallie. 1993. *The Body of God: An Ecological Theology*. Fortress Press.

Rahman, Sira Abdul. 2017. "Religion and Animal Welfare — An Islamic Perspective." *Animals* 7 (2): 11.

Reggente, Melissa A.L.V., Elena Papale, Niall McGinty, Lavinia Eddy, Giuseppe Andrea De Lucia, and Chiara Giulia Bertulli. 2018. "Social Relationships and Death-Related Behaviour in Aquatic Mammals: A Systematic Review." *Philosophical Transactions of the Royal Society B: Biological Sciences* 373 (1754): 20170260.

Weatherdon, Meaghan S. 2022. "Religion, Animals, and Indigenous Traditions." *Religions* 13 (7): 654.

White, Lynn, Jr. 1967. "4. The Historical Roots of Our Ecologic Crisis." *Science* 155 (3767): 1203–1207.

Restoration Ecology

Restoration ecology is "the study of the relationships among organisms and their environment in a restoration context ... (it) draws on all branches of ecological science and spans genes to entire landscapes" (Palmer et al. 2016, 3). The field of restoration ecology is focused on how to repair damage to ecological systems caused by humans and how to protect existing healthy ecosystems using an interdisciplinary approach that includes ecology, environmental science, economics, sociology, and politics (Cairns and Heckman 1996).

In simple terms, restoration ecology is the study of ecological restoration and is either goal-oriented (returning an ecosystem to close to its condition prior to disturbance) or process-oriented (repairing damage caused by humans to the diversity and dynamics of ecosystems) (Cairns and Heckman 1996). The goal-oriented definition was developed in 1992 by the US National Research Council (NRC) and provides objective research into ecological restoration designs. The process-oriented definition shifts the emphasis to taking the necessary actions to return the ecosystem to a natural ecological state, including a judgment of the need for restoration, goal setting, and evaluation (→ Applied Zooarchaeology) (Cairns and Heckman 1996, 170).

There are many examples of ecological restoration, such as rehabilitating forests and grasslands, erosion control, removing invasive species, reintroducing lost species, restoring coastal habitats and coral reefs, and creating wildlife corridors (→ Invasive Animal Species) (→ Rewilding) (→ Empty Forest) (→ Defaunation) (→ Animal Forests) (→ Ecological Connectivity). There are eight principles that underpin ecological restoration: 1) engage stakeholders, 2) draw on multiple types of knowledge, 3) be informed by native ecosystems while considering environmental change, 4) support ecosystem recovery, 5) assess against clear goals and objectives using measurable indicators, 6) seek the highest level of recovery attainable, 7) plan for gaining cumulative value when applied at large scales, and 8) engage as part of a continuum of restoration activities (Reaser et al. 2021). In a meta-analysis of 89 assessments of restorations in a wide range of ecosystem types across the globe, Benayas et al (2009) found that ecological restoration increased provision of biodiversity by 44 percent and ecosystem services by 25 percent, indicating that ecological restoration has benefits for biodiversity conservation and supporting human livelihoods (→ Biodiversity) (→ Ecosystem Services). For example, the restoration of pine-grassland communities in the Ouachita National Forest in Arkansas has been beneficial to the endangered Eastern Wood-Peewee as well as the declining red-cockaded woodpecker (Cairns and Heckman 1996).

The overall goal of restoration is to increase the resilience of ecosystems and their ability to function in a changing environment (→ Global Environmental Change). This goal argues for a focus on the causes of environmental problems and what we can do to mitigate them rather than on the symptoms such as loss of species or scarcity of water (Palmer and Stewart 2020), emphasizing that process-oriented restoration actions are key to achieving ecosystem sustainability.

References

Benayas, José M. Rey, Adrian C. Newton, Anita Diaz, and James M. Bullock. 2009. "Enhancement of Biodiversity and Ecosystem Services by Ecological Restoration: A Meta-Analysis." *Science* 325 (5944): 1121–1124.

Cairns, John, Jr., and John R. Heckman. 1996. "Restoration Ecology: The State of an Emerging Field." *Annual Review of Energy and the Environment* 21: 167–189.

Palmer, Margaret, and Graham A. Stewart. 2020. "Ecosystem Restoration is Risky … but We Can Change That." *One Earth* 3: 661–664.

Palmer, Margaret A., Joy B. Zedler, and Donald A. Falk. 2016. "Ecological Theory and Restoration Ecology." In *Foundations of Restoration Ecology*, 2nd ed., edited by Margaret A. Palmer, Joy B. Zedler, and Donald A. Falk, 3–26. Island Press.

Reaser, Jamie K., Arne Witt, Gary M. Tabor, Peter J. Hudson, and Raina K. Plowright. 2021. "Ecological Countermeasures for Preventing Zoonotic Disease Outbreaks: When Ecological Restoration is a Human Health Imperative." *Restoration Ecology* 29 (4): e13357.

Rewilding

Originally called "wilderness recovery" (Noss 1985), rewilding is the "process of rebuilding, following major human disturbance, a natural ecosystem by restoring natural processes and the complete or near complete food web at all trophic levels as a self-sustaining and resilient ecosystem with biota that would have been present had the disturbance not occurred" (Carver et al. 2021, 1888). The goal of rewilding is the restoration of native ecosystems that contain the full range of species at all trophic levels (not just the protection of single species and habitats) (→ Biodiversity); rewilded ecosystems are dynamic and self-sustaining that require no (or minimal) intervention (e.g., *natura naturans* or nature doing what nature does) (Carver et al. 2021). The vision of rewilding is to reach "a healthier balance between Nature and human society – one that grounds people in Nature as much as it sustains the actual ground of the natural world" (→ One Health) (Soulé 1999, 39).

There are 10 basic principles of successful rewilding (Carver et al. 2021). Rewilding 1) uses wildlife to restore trophic interactions, and keystone species who maintain the ecosystem should be reintroduced or reinforced (→ Wilderness Animals) (→ Keystone Species) (→ Trophic Cascades); 2) employs landscape planning that considers core areas, connectivity, and co-existence between wildlife, humans, and livestock (→ Livestock Production) (→ Ecological Connectivity) (→ Coexistence); 3) focuses on the recovery of ecological processes, interactions, and conditions (→ Restoration Ecology) based on reference ecosystems, including the replacement of extinct species with suitable wild surrogates; 4) recognizes that ecosystems are dynamic and constantly changing; 5) anticipates the effects of climate change (→ Climate Change) and where possible mitigates impacts such as providing connectivity for species movements (→ Ecological Connectivity); 6) requires local engagement and support; 7) is informed by both science and Indigenous and local knowledge (→ Citizen Science) (→ Indigenous and Decolonial Perspectives); 8) is adaptive and dependent on monitoring and feedback; 9) recognizes the intrinsic value of all species and ecosystems (→ Ecocentrism); and 10) requires a paradigm shift in the coexistence of humans and nature (→ One Health).

There have been numerous successful terrestrial rewilding projects. The restoration of wolves in the northern Yellowstone ecosystem triggered a tri-trophic cascade involving predators, prey, and plants. The presence of wolves caused a decrease in elk populations that in turn caused an increase in woody browse plant species and canopy cover, and both beaver and bison numbers increased due to the increase in plants and herbaceous forage resulting from less competition with elk (Ripple and Beschta 2012). Researchers have documented that rewilding in Europe restores biodiversity quickly and at scale (Hart et al. 2023). For example, the Chornobyl Exclusion Zone has undergone trophic rewilding since the 1986 nuclear disaster with the introduction of Przewalski's horses and European bison to the site, and wolves, moose, deer, and wild boar have naturally recolonized the area and grown to population sizes similar to (and higher in the case of wolves) than those in other nature reserves in the region. There have also been dramatic habitat changes in the exclusion zone with an increase of wetland avian predators, as well as recolonization by species extirpated prior to 1986, such as the white-tailed eagle and the greater spotted eagle (Hart et al. 2023).

Marine rewilding is "a collaborative, human-initiated, nature-led process involving a range of approaches that work systemically to make degraded marine and coastal environments more ecologically complex while supporting greater social inclusion and enabling sustainable economic and cultural opportunities" (Brooker et al. 2025, 3–4). This definition places greater emphasis on non-ecological aspects, such as social values, and reflects the potential for rewilding to result in an alternative ecological state rather than an original state (Brooker et al. 2025). While there are many differences between terrestrial and ocean areas (such as the ocean's high connectivity, openness, national jurisdiction, and scarcity of scientific knowledge), the main difference is the absence of an ecological reference point or baseline in marine rewilding (→ Applied Zooarchaeology) (Cornerier 2024). In addition, compared to terrestrial areas that can be owned, marine areas are open with a greater density of stakeholder activities at sea, such as fisheries, aquaculture farms, and ecotourism (→ Aquaculture) (→ Ecotourism) (Cornerier 2024).

Marine Rewilding Initiatives (MRIs) (also called reforesting the ocean, seawilding, rewilding the sea, and seaforestation) have gained momentum only in the last decade, but they are growing throughout the world (Cornerier 2024). The reintroduction or translocation of keystone marine species is part of the early stages of marine rewilding that focuses on bottom-up habitat-constructing species (such as oysters, corals, and mangroves), and top-down trophic influencers (such as shark and whale species) (→ Animal Forests) (→ Keystone Species) (→ Trophic Cascades) (Brooker et al. 2025). Active rewilding has been initiated in the marine environment, such as the reintroduction of a captive-bred apex predator, the zebra shark, to the Raja Ampat Islands coastal area in 2020 (Cornerier 2024). This is an important initiative since one of the main challenges facing marine rewilding is the limit of scientific knowledge on how to rewild the sea through the reintroduction of wild fauna and flora (→ Applied Zooarchaeology), particularly limited breeding techniques (Cornerier 2024). The restoration of degraded ecosystems and the return of marine biodiversity through rewilding efforts will encourage not only a healthier ocean but the sustainability of the entire planet.

References

Brooker, Esther E., Gerald Midgley, Neil Burns, Charlotte E. Trotman, Amanda Gregory, and Charlotte Rachael Hopkins. 2025. "Defining Marine Rewilding Can Help Guide Theory and Practice in Marine Conservation." *Communications Earth & Environment* 6: 241. https://doi.org/10.1038/s43247-025-02155-x.

Carver, Steve, Ian Convery, Sally Hawkins, et al. 2021. "Guiding Principles for Rewilding." *Conservation Biology* 35: 1882–1893. https://doi.org/10.1111/cobi.13730.

Cornerier, Ariane. 2024. "Thinking Marine Rewilding: Adapting a Terrestrial Notion to the Marine Realm: Definition, Practices and Theories of Marine Rewilding." *European Journal of Geography: Environment, Nature, Landscape* 1071. https://doi.org/10.4000/122yc.

Hart, Emma E., Amy Haigh, and Simone Ciuti. 2023. "A Scoping Review of the Scientific Evidence Base for Rewilding in Europe." *Biological Conservation* 285: 110243. https://doi.org/10.1016/j.biocon.2023.110243.

Noss, Reed F. 1985. "Wilderness Recovery and Ecological Restoration: An Example for Florida." *Earth First!* 5 (8): 18–19.

Ripple, William J., and Robert L. Beschta. 2012. "Trophic Cascades in Yellowstone: The First 15 Years Wolf Reintroduction." *Biological Conservation* 145: 205–213.

Soulé, Michael E. 1999. "An Unflinching Vision: Networks of People for Networks of Wildlands." *Wild Earth* 9 (4): 38–46.

Rights

Rights refer to justified claims or entitlements that outline the protections and obligations individuals hold in society. They function as normative principles that specify what individuals are permitted to do and what they are owed legally, ethically, and socially. Rights historically center on humans; however, scholars are increasingly interested in the extension of rights to non-human animals so that they are not treated merely as property (→ Property) (Regan and Singer 1989; Francione 2008; Favre 2017). This work is based on the argument that non-human animals are sentient beings who experience various emotional states and can form social bonds (→ Sentience) (→ Cognition)

(→ Empathy). Therefore, they could hold claims against harmful actions to them and their species by humans. Legal and political theorists have since explored how these claims might translate into law and policy, including debates over animal personhood, welfare versus abolitionist approaches, and the limits of existing animal protection statues such as the Animal Welfare Act and the Marine Mammal Protection Act (→ Animal Law) (→ Welfare) (→ Abolitionism) (→ Animal Welfare Act) (→ Marine Mammal Protection Act).

Central to the debate on rights is the idea of legal personhood being extended to non-human animals, particularly those shown to have cognitive complexity (→ Cognition). Over the past two decades, advocates have petitioned the courts to consider the personhood of great apes and elephants (→ Advocacy) (Wise 2000; Andrews 2020). For example, in a landmark legal case, the Nonhuman Rights Project (NhRP), an animal rights organization, filed a series of petitions demanding the extension of "legal personhood" be granted to Happy, a female Asian elephant who has resided at the Bronx Zoo in New York City since the 1970s. NhRP asserted that Happy should be recognized as a legal person and be given "bodily liberty" to retire to an elephant sanctuary. In 2022, the New York Court of Appeals rejected the claim in a 5–2 decision, asserting that habeas corpus does not apply to non-human animals. Although the petition was denied, the case advanced public awareness and debate about animal rights, personhood status, and ethics regarding captive wild animals (→ Captive Wild Animals). The discussion is global and some countries such as Bolivia, Ecuador, and New Zealand, have granted legal rights to non-humans (→ Ecodemocracy).

Understanding the rights of non-human animals has direct implications for sustainability, particularly in shaping legal and regulatory approaches to wildlife crime, including how exploitation, illegal trade, and harm to animals are defined, governed, and addressed within broader environmental protection frameworks (→ Wildlife Crime) (→ Wildlife Trade) (→ Convention on International Trade in Endangered Species). Industrial systems that treat animals purely as objects of profit, such as intensive livestock production, negatively impact the wellbeing of agricultural animals, wild animals, ecosystems, and humans (→ Ecosystems) (→ Livestock Production) (→ Wilderness Animals). A rights-based framework encourages society to question and reconsider industrial animal production and other use-based practices as potential violations of the rights of non-human animals in support of more ecologically responsible, humane, and sustainable practices.

References

Andrews, Kristin. 2020. *The Animal Mind: An Introduction to the Philosophy of Animal Cognition*. Routledge.

Favre, David. 2017. "Animals as Living Property." In *The Oxford Handbook of Animal Studies*, edited by Linda Kalof, 65–80. Oxford University Press.

Francione, Gary Lawrence. 2008. *Animals as Persons: Essays on the Abolition of Animal Exploitation*. Columbia University Press.

Regan, Tom, and Peter Singer. 1989. *Animal Rights and Human Obligations.* Prentice Hall.
Wise, Steven. 2000. *Rattling the Cage: Toward Legal Rights for Animals.* Da Capo Press.

Road Ecology

Road ecology is the scientific study of the impacts of roads, road systems, and traffic on: 1) air, water, and soil pollution; 2) barrier and mortality impacts to wildlife (plants and animals) movement; 3) noise and light emissions to nearby human and wildlife habitats; 4) changes to natural geomorphic and hydrologic attributes and processes; 5) invasion of non-native and damaging species and extraction activities; and 6) other modifications of ecosystem attributes and processes (Shilling 2025). Of particular importance is how non-human animals and the environment are affected by roads, highways, railroad track beds, and other transportation infrastructure (Coffin et al. 2021). These corridors lead to land use change, block animal movement, fragment ecosystems, and genetically isolate animal populations (→ Ecological Connectivity) (→ Habitat Loss) (Balčiauskas et al. 2025).

The most dramatic impact studies in road ecology are of wildlife killed on roads (roadkill). While the numbers globally are hard to estimate, data for Europe suggest that 194 million birds and 29 million mammals are killed per year on roads (Groot Bruinderink and Hazebroek 1996). Humans are also impacted — in Europe collisions with ungulates alone account for 200 human fatalities and 29,000 human injuries per year. Many of these accidents kill relatively common species such as deer and moose who share habitat with humans (→ Liminal Animals). However, threatened and endangered species are also harmed, and roadkill can have a significant impact on species survival (→ Endangered and Threatened Species). For some species such as the Iberian lynx and African wild dog roadkill is one of the largest causes of their mortality.

More sustainable road system designs, such as incorporating safe wildlife crossings that overpass or underpass roads and other infrastructure, will reduce the adverse effects of road systems on non-human animals and help preserve biodiversity (→ Animal-Inclusive Design) (→ Ecological Connectivity). In addition, ledges and vegetation cover help increase movement for small mammals, and canopy bridges and vegetated medians improve road crossing for arboreal mammals further minimizing biodiversity loss (→ Ecotones) (Soanes et al. 2024).

References

Balčiauskas, Linas, Andrius Kučas, and Laima Balčiauskienė. 2025. "A Review of Wildlife–Vehicle Collisions: A Multidisciplinary Path to Sustainable Transportation and Wildlife Protection." *Sustainability* 17 (10): 4644. https://doi.org/10.3390/su17104644.
Coffin, Alisa W., Douglas S. Ouren, Neil D. Bettez, et al. 2021. "The Ecology of Rural Roads: Effects, Management, and Research." *Issues in Ecology, Report No. 23. Ecological Society of America.* https://research.fs.usda.gov/treesearch/68966.

Groot Bruinderink, G.W.T.A., and Eric J. Hazebroek. 1996. "Ungulate Traffic Collisions in Europe." *Conservation Biology* 10 (4): 1059–1067.

Shilling, Fraser. 2025. "What Is Road Ecology?" Road Ecology Center, UC Davis. https://roadecology.ucdavis.edu/news/what-road-ecology.

Soanes, Kylie, Trina Rytwinski, Lenore Fahrig, et al. 2024. "Do Wildlife Crossing Structures Mitigate the Barrier Effect of Roads on Animal Movement? A Global Assessment." *Journal of Applied Ecology* 61 (3): 417–430. https://doi.org/10.1111/1365-2664.14582.

S

Sanctuaries

Animal sanctuaries are places that give refuge and safety to non-human animals. Non-human animals may become residents in sanctuaries after being rescued from abuse, neglect, exploitation, or displacement. Unlike shelters or zoos, sanctuaries do not breed, buy, sell, or use animals for entertainment or commercial gain (→ Zoos and Aquariums). The expectation when a non-human animal is admitted into a sanctuary is that they will reside under the sanctuary's protection and care for the remainder of their life. Sanctuaries differ widely in their size, organizational structure, and mission, and they accommodate a diverse range of species, from farmed and companion animals to wildlife and exotics (→ Domestic Animals) (→ Wilderness Animals).

Farm Sanctuary, established in 1986 in upstate New York and then expanded into California in the US, is often cited as the first large-scale farmed animal sanctuary and a catalyst for a broader movement linking rescue, education, and advocacy (→ Advocacy). It was founded by Gene Baur and Lorri Houston after rescuing a sheep named Hilda who had been discarded still alive on a pile of dead animals. Gene Baur helped to popularize the idea that sanctuaries can function as both safe havens for non-human animals and political sites for education and advocacy (Baur 2008; Baur and Stone 2015). Historically, informal sanctuaries emerge as people choose to care for injured, abandoned, or unwanted animals, often with limited resources and little formal training. Many of these efforts depend heavily on private donations that create financial instability and can constrain their ability to provide consistent, long-term care. The absence of standardized practices in animal husbandry, combined with the emotional and physical demands of caring for chronically vulnerable animals, has contributed to high turnover and organizational fragility across the sector. As a result, the sanctuary landscape has long been marked by uneven quality, with some facilities offering exemplary care and others struggling to meet basic welfare standards. Scholars such as Sue Donaldson and Will Kymlicka (2015) argue that when run well, sanctuaries not only provide refuge for non-human animals but also offer space for intentional

DOI: 10.4324/9781003528142-19

interspecies communities to develop with the potential to break exploitative social norms (→ Multispecies Futures).

By removing animals from abusive situations or exploitative industries, many sanctuaries make public the stories of individual animals and denounce factory farming and commercial wildlife displays that can be used to highlight the link between animal exploitation, land use, and climate change. Sanctuary-based education programs, visits, and online storytelling can foster empathy for animals and encourage plant-based diets, reduce consumption of animal products, and encourage broader awareness of ecological systems (→ Naturalist Programs) (→ Empathy) (→ Vegetarianism) (→ Veganism) (→ Ecosystems) (Speiran 2025). At their best, animal sanctuaries function as living laboratories for low-carbon, care-centered ways of relating to animals and landscapes.

References

Baur, Gene. 2008. *Farm Sanctuary: Changing Hearts and Minds about Animals and Food*. Simon and Schuster.

Baur, Gene, and Gene Stone. 2015. *Living the Farm Sanctuary Life: The Ultimate Guide to Eating Mindfully, Living Longer, and Feeling Better Every Day*. Rodale.

Donaldson, Sue, and Will Kymlicka. 2015. "Farmed Animal Sanctuaries: The Heart of the Movement?" *Politics and Animals* 1: 50–74.

Speiran, Siobhan I.M. 2025. "The 'Sanctuary Gap': Reviewing the Research on Captive Wildlife Sanctuary Tourism." *Animals* 15 (4): 496.

Sentience

Sentience refers to an animal's capacity to consciously perceive and experience the world. This includes the ability to feel pain, pleasure, fear, and other affective states. Scientific research in neuroscience, ethology, and cognitive biology increasingly support the view that many, if not all, non-human animals possess some level of sentience (Browning and Birch 2022; de Souza Valente 2025). Many ethical theories afford consideration based on the ability to suffer or to flourish, so the realization that many species are sentient extends ethical consideration to them under those theories (→ Utilitarianism) (→ Rights) (→ Cognition) (→ Capabilities Ethics) (→ Ecofeminism).

There are many examples of studies exploring the sentience of non-human animal species. Cephalopods (cuttlefish, squid, octopus, and nautilus) have sophisticated cognitive capabilities, including problem-solving abilities, tool use, and exploratory curiosity (→ Cognition) (Birch et al. 2021; Browning 2019). They also exhibit stress responses, have individual personality traits, and experience pain (Godfrey-Smith 2016). Other invertebrates are sentient as well, including crustaceans, insects, arachnids, and bivalves (de Souza Valente, 2025).

As sentience is documented among more species, including invertebrates, policy debates arise. For instance, in 2021, after an extensive review of scientific studies on cephalopod sentience by the London School of Economics, the Animal Welfare Bill in the United Kingdom (UK) was amended to include cephalopods as sentient beings. The review concluded that cephalopods are capable of pain, learning, and complex behaviors. The amended act means that the government must now consider the welfare needs of cephalopods not just in scientific research, but more broadly (Osorio 2023). Prior to this amendment in the UK, cephalopods had limited protections as research subjects under the animal portion of the Scientific Procedures Act of 1986. The amended law does not change fishing, aquaculture, or restaurant industry practices, nor does it immediately ban practices like boiling or slicing cephalopods while alive, but these practices are likely to be considered inhumane in the future. Although the immediate changes are limited, this legislation marked a significant shift in the inclusions of invertebrates in welfare debate and policy (→ Welfare).

Understanding animal sentience has direct implications for sustainability. Sustainable systems aim to balance ecological integrity, human well-being, and ethical responsibilities toward other species. Recognizing animals as sentient challenges models of production and resource extraction that rely on the intensive use of animals, encouraging shifts toward practices that minimize suffering, reduce ecological harm, and support more resilient ecosystems. Incorporating sentience into sustainability frameworks helps promote forms of agriculture, fisheries, and wildlife management that consider both environmental limits and the lived experiences of animals, fostering more humane and ecologically aligned socio-ecological futures.

References

Birch, Jonathan, Charolotte Burn, Alexandra Schnell, Heather Browning, and Andrew Crump. 2021. "Review of the Evidence of Sentience in Cephalopod Molluscs and Decapod Crustaceans." LSE Consulting. The London School of Economics and Political Science. www.lse.ac.uk/News/News-Assets/PDFs/2021/Sentience-in-Cephalopod-Molluscs-and-Decapod-Crustaceans-Final-Report-November-2021.pdf.

Browning, Heather. 2019. "What Is Good for an Octopus?" *Animal Sentience* 4 (26): 1–3.

Browning, Heather, and Jonathan Birch. 2022. "Animal Sentience." *Philosophy Compass* 17 (5): e12822.

de Souza Valente, Cecília. 2025. "Rethinking Sentience: Invertebrates as Worthy of Moral Consideration." *Journal of Agricultural and Environmental Ethics* 38 (1): 3. https://doi.org/10.1007/s10806-024-09940-2.

Godfrey-Smith, Peter. 2016. *Other Minds: The Octopus, the Sea, and the Deep Origins of Consciousness*. Farrar, Straus and Giroux.

Osorio, Daniel. 2023. "Cephalopods and the Law." *Current Biology* 33 (20): R1078–R1080.

Sentinels

Animal sentinels are non-human species whose health, behavior, or population dynamics provide early warning signs of environmental contamination that may be a threat to human populations. Non-human animals are often more sensitive to environmental hazards due to distinctive biological and ecological traits, such as elevated metabolic rates, higher respiration, and feeding behaviors that facilitate the accumulation of toxins at concentrations exceeding those found in humans. When animals are used to detect environmental toxins or when they are monitored in their natural habitats for toxin exposure, they become animal sentinels.

The most iconic animal sentinel is the "canary in the coal mine." Starting in the early twentieth century, miners used canaries underground to monitor gases like carbon monoxide that could pose a risk to human health. Canaries were historically selected as sentinel species because their rapid metabolic rate, heightened sensitivity to carbon monoxide, low cost of breeding, and ease of domestication made them effective indicators of toxic gas exposure (→ Domestic Animals) (Pollock 2016; Burrell 1914). Even before canaries were introduced in mining, historical records indicate that non-human animals were frequently employed as sentinels to evaluate the safety of environments and products for humans. Examples include dogs trained to detect smoke or spoiled food and livestock observed for signs of water contamination (Rabinowitz et al. 2005).

The use of non-human animals as environmental risk indicators occurs in two ways: intentional or unintentional. Intentional animal sentinels are animals who are deliberately put into environments to assess immediate toxin risk to humans, such as the canary in the coal mine. Alternatively, animals residing in environments with historic contamination or contamination due to the onset of a disaster can be monitored to assess the long-term impact of exposure and potential risk to human health (→ Disasters) (Whitley 2018; Mattes and Whitley 2021). These animals are unintentional sentinels. Researchers are increasingly moving beyond the use of intentional sentinels, instead monitoring animals in natural habitats to assess contamination and the potential impact to humans, such as the monitoring of mollusks and fish to investigate aquatic contamination and the use of domestic animals to determine exposure to lead, pesticides, or industrial chemicals in shared spaces with humans (→ Animal Research) (Rabinowitz et al. 2009). Ecologists have also started to use the term "ecosystem sentinels," for apex predators or migratory species whose behaviors can indicate large-scale environmental shifts or ecological collapse (→ Learning for Survival) (Hazen et al. 2024). Ecosystem sentinels are inherently unintentional sentinels.

Animal sentinels show us how non-human animals experience environmental toxins and signal ecological risks before they appear in human populations (National Research Council Committee on Animals as

Monitors of Environmental Hazards 1991). From an intersectional environmental justice perspective, animal sentinels illuminate how ecological harm and social inequality are intertwined (→ Environmental Justice). Given that underrepresented and marginalized communities, particularly low-income neighborhoods and communities of color, are more likely to be situated near industrial facilities, waste sites, and polluted waterways, unintentional animal sentinels in these environments reveal patterns of toxic exposure that align with human vulnerability and the intersection of marginalized identities. Recognizing the value of animal sentinels advances both scientific monitoring and ethical accountability by linking biodiversity conservation, and ecological health to the broader pursuit of intersectional environmental justice and long-term sustainability (→ Biodiversity) (→ Intersectional Environmentalism) (→ Environmental Justice).

References

Burrell, George Arthur. 1914. *The Use of Mice and Birds for Detecting Carbon Monoxide after Mine Fires and Explosions*. 3rd ed. US Dept. of the Interior, Bureau of Mines.

Hazen, Elliott L., Matthew S. Savoca, T.J. Clark-Wolf, Max Czapanskiy, Peter M. Rabinowitz, and Briana Abrahms. 2024. "Ecosystem Sentinels as Early-Warning Indicators in the Anthropocene." *Annual Review of Environment and Resources* 49 (1): 573–598.

Mattes, Seven, and Cameron T. Whitley. 2021. "Entangled Impacts: Human-Animal Relationships and Energy Development." In *Energy Impacts: A Multidisciplinary Exploration of North American Energy Development*, edited by Jeffrey B. Jacquet, Julia H. Haggerty, and Gene L. Theodori. University Press of Colorado.

National Research Council Committee on Animals as Monitors of Environmental Hazards. 1991. *Animals as Sentinels of Environmental Health Hazards*. National Academies Press.

Pollock, Christal. 2016. "The Canary in the Coal Mine." *Journal of Avian Medicine and Surgery* 30 (4): 386–391.

Rabinowitz, Peter, Matthew Scotch, and Lisa Conti. 2009. "Human and Animal Sentinels for Shared Health Risks." *Veterinaria Italiana* 45 (1): 23.

Rabinowitz, Peter M., Zimra Gordon, Rebecca Holmes, et al. 2005. "Animals as Sentinels of Human Environmental Health Hazards: An Evidence-Based Analysis." *EcoHealth* 2 (1): 26–37.

Whitley, Cameron T. 2018. "Engines, Sentinels and Objects: Assessing the Impacts of Unconventional Energy Development on Animals in the Marcellus Shale Region." In *Fractured Communities: Risk, Impacts, and Protest Against Hydraulic Fracking in the U.S. Shale Regions*, edited by Anthony Ladd. Rutgers University Press.

Speciesism

Speciesism is the assumption that humans are inherently more valuable than non-human animal species and therefore humans are entitled to use non-human animals for human purposes without serious moral consideration

(→ Anthropocentrism) (Horta 2010). The term was first used in 1970 by psychologist and animal rights advocate Richard Ryder and then popularized by philosopher Peter Singer in *Animal Liberation* (1975) (→ Rights). It continues to be used to highlight how cultural norms, legal systems, and economic structures routinely privilege human interests while minimizing or disregarding the experiences, needs, and welfare of non-human animals (→ Welfare). Contemporary scholars emphasize that speciesism functions as a social hierarchy like racism, sexism, or homophobia where moral value is assigned based by species instead by sentience, ecological role, or capacity to suffer (→ Sentience) (Caviola et al. 2019).

Understanding speciesism is important for analyzing how human societies justify practices that cause widespread harm to animals despite growing scientific evidence of animal cognition, emotion, and social complexity. Livestock production is one of these cases. Although cows, pigs, and chickens are capable of pain, social bonding, and sophisticated forms of communication, they are typically excluded from legal protections and rights afforded to companion animals or endangered species (→ Animal Languages) (→ Livestock Production) (→ Animal Welfare Act). Industrial agriculture relies on confinement systems that prioritize efficiency and production over behavioral needs, even when evidence shows that such systems reduce welfare and increase stress-related illness. At the same time, many societies view dogs and cats as family members (Pickersgill et al. 2023), granting them rights and protections unavailable to farmed animals (→ Rights). This contrast demonstrates how speciesism shapes moral judgement where animals considered useful for food or labor are often seen as less worthy of care, while animals valued for companionship receive more empathetic treatment (→ Empathy).

Speciesism also has significant implications for sustainability. Livestock production is a major driver of deforestation, greenhouse gas emissions, and biodiversity loss, yet these ecological impacts persist partly because speciesist assumptions normalize large-scale exploitation of animals and their habitats (→ Biodiversity) (→ Habitat Loss). By framing certain species primarily as resources, society overlooks the roles animals play in maintaining healthy ecosystems. Addressing sustainability therefore requires rethinking human-animal relationships, recognizing animals as ecological partners rather than commodities, and integrating animal welfare considerations into environmental policy. Reducing speciesism can promote more equitable and ecologically grounded approaches to conservation, food systems, and climate mitigation.

References

Caviola, Lucius, Jim A.C. Everett, and Nadira S. Faber. 2019. "The Moral Standing of Animals: Towards a Psychology of Speciesism." *Journal of Personality and Social Psychology* 116 (6): 1011.

Horta, Oscar. 2010. "What Is Speciesism?" *Journal of Agricultural and Environmental Ethics* 23 (3): 243–266.
Pickersgill, Olivia, Daniel S. Mills, and Kun Guo. 2023. "Owners' Beliefs Regarding the Emotional Capabilities of Their Dogs and Cats." *Animals* 13 (5): 820.
Singer, Peter. 1975. *Animal Liberation. Towards an End to Man's Inhumanity to Animals*. Granada Publishing Ltd.

Sustainable Development Goals (SDGs)

In 2015, the United Nations, as a result of prompting by Colombia, Guatemala and Peru, adopted 17 Sustainable Development Goals (SDGs) with 169 carefully defined targets for meeting those goals (→ Sustainability) (Santos 2023; Kamau et al. 2018). The SDGs and the targets for each were developed based on broad international input and were intended to set specific targets for improving human well-being and reducing harm to the environment. The hope was to achieve the goals by a target date of 2030. Two goals, SDG 14 "Life Under Water" and SDG 15 "Life on Land" engage the well-being of non-human animals. But overall, the goals are anthropocentric, with most goals specifically about human well-being, and even SDG 14 and 15 emphasize their contribution to human well-being (→ Anthropocentrism). Other goals, such as SDG 12 "Responsible Consumption and Production" and SDG 13 "Climate Action" have substantial impact on non-human animals. SDG 15 is now explicitly linked to the 30x30 pledge to protect 30 percent of the Earth's land area by 2030. Such expansion in protected areas would help preserve biodiversity and prevent extinctions (→ Biodiversity) (→ Extinction) (Convention on Biodiversity, n.d.).

There are 10 indicators and 10 targets for SDG 14 (Life Under Water). While all have a potential impact on marine animals, five are specifically targeted at fisheries. There are 14 indicators and targets for SDG 15 (Life on Land). Two focus on animal trafficking (→ Wildlife Trade) (→ Wildlife Crime). One deals with endangered species, specifically those on the "Red List" (→ Convention on International Trade in Endangered Species). Five deal explicitly with biodiversity and one with invasive species (→ Biodiversity) (→ Invasive Species).

The most recent reports on progress toward the SDGs have raised concern that progress on many goals is lagging and the 2030 targets will not be met (Xing et al. 2025; Zhang et al. 2023; Xu et al. 2020). It appears that more progress is being made on "Life on Land" than on "Life Under Water" but there is huge variation in progress across countries. The UN rates countries on what percentage of each goal has been achieved, with an aggregate score from 0-100 where 0 would mean no progress on any goal and 100 would mean fulfillment of all goals (United Nations 2025). By 2025, Finland, Sweden, and Denmark were all above 85, with 19 other European countries and Japan above 80. Canada was at 79, the US at 75. The lowest scoring countries (below 50) tended to be either low-income countries in Africa and/or countries with a recent history of violent conflict. Data were not available for 26 of the 197 countries who have signed on to the goals. For some metrics, there has been either no progress or backsliding leading to calls for renewed efforts and an acknowledgement that having specific goals and metrics to measure progress can be the basis for further evaluation and new efforts (United Nations 2025). The SDGs may be one of only a handful of examples of sustainability considerations involving animals being systematically quantified and monitored.

References

Convention on Biodiversity. n.d. "*Target 3.*" Secretariat of the Convention on Biological Diversity. Accessed February 18, 2026. www.cbd.int/gbf/targets/3.

Kamau, Macharia, Pamela Chasek, and David O'Connor. 2018. *Transforming Multilateral Diplomacy: The Inside Story of the Sustainable Development Goals*, 1st ed. Routledge. https://doi.org/10.4324/9780429491276.

Santos, Juan Manuel. 2023. "The Sustainable Development Goals and the Multidimensional Nature of Poverty." In *The Battle Against Poverty*, 1st ed., edited by Juan Manuel Santos. Oxford University Press. https://doi.org/10.1093/oso/9780192885234.003.0009.

United Nations. 2025. *The Sustainable Development Goals Report 2025*. United Nations.

Xing, Qiang, Linlin Lu, Li Wang, et al. 2025. "Country-Specific Progress toward the Sustainable Development Goals: Past, Present, and Prospects." *Proceedings of the National Academy of Sciences* 122 (52): e2524299122. https://doi.org/10.1073/pnas.2524299122.

Xu, Zhenci, Sophia Chau, Xiuzhi Chen, et al. 2020. "Assessing Progress towards Sustainable Development over Space and Time." *Nature* 577 (2 January 2020): 74–78. https://doi.org/10.1038/s41586-019-1846-3.

Zhang, Yuqian, Yingjie Li, and Jianguo Liu. 2023. "Global Decadal Assessment of Life below Water and on Land." *iScience* 26 (4): 106420. https://doi.org/10.1016/j.isci.2023.106420.

Sustainability

The concept of sustainability has deep historical roots, but its current use is usually traced to *Our Common Future* where sustainable development was defined as "development that meets the needs of the present without compromising the ability of future generations to meet their own needs" (World Commission on Environment and Development 1987, 43; Caradonna 2022). The focus on the future leads to consideration of environmental change and especially the depletion of resources that might be needed for later generations. In its original formulation, the idea is anthropocentric in that the well-being of other species is considered only insofar as it matters for the well-being of contemporary humans and future generations of humans (→ Anthropocentrism). Some discussions continue to use the term "sustainable development" to link to traditional arguments for the need for economic development to improve the conditions of the least well-off humans. But other discussions use the term "sustainability" to allow for, but not give primary emphasis, to economic growth.

A focus on future generations sometimes leads to very technical discussions about how to value the future (Neumayer 2025). As a result, some have offered a simpler and less anthropocentric approach, viewing sustainability as improving the well-being of humans and other species while reducing stress on the environment (Dietz 2023). Others have expanded the multi-generational concern to "multispecies sustainability" to emphasize consideration all living species (such as a tree and the multitude of species interacting with and dependent on it, from plants and microbes to insects and birds) (→ Multispecies Futures) (→ Biocentrism) (Rupprecht et al. 2020).

Many national governments and the global community have embraced the general idea of sustainability. For example, in 2015 the United Nations adopted 17 Sustainable Development Goals with 169 carefully defined targets for meeting those goals (→ Sustainable Development Goals) (Santos 2023). Of the 17, two, "Life Below Water" and "Life on Land," specifically take account of the well-being of other species, and several others have implications for animals. But the broader, multi-species approaches to sustainability argue for the consideration of all species in sustainability analysis.

References

Caradonna, Jeremy L. 2022. *Sustainability: A History*, Revised and Updated edition. Oxford University Press.

Dietz, Thomas. 2023. *Decisions for Sustainability: Facts and Values*. Cambridge University Press.

Neumayer, Eric. 2025. *Weak versus Strong Sustainability: Exploring the Limits of Two Opposing Paradigms*, 5th ed. Edward Elgar Publishing.

Rupprecht, Christoph D.D., Joost Vervoort, Chris Berthelsen, et al. 2020. "Multispecies Sustainability." *Global Sustainability* 3: e34. https://doi.org/10.1017/sus.2020.28.

Santos, Juan Manuel. 2023. "The Sustainable Development Goals and the Multidimensional Nature of Poverty." In *The Battle Against Poverty*, 1st ed., edited by Juan Manuel Santos, 132–147. Oxford Academic. https://doi.org/10.1093/oso/9780192885234.003.0009.

World Commission on Environment and Development. 1987. *Our Common Future*. Oxford University Press. https://digitallibrary.un.org/record/139811?v=pdf.

T

Telecoupling

Telecoupling refers to the connections across human and ecological systems over distances (Liu et al. 2013). It includes the exchange of energy, matter, and information among human and natural systems across spatial, temporal, and organizational borders. Examples include tourism, trade, migration, species invasion, pollution, and flows and use of ecosystem services and goods across boundaries. Such connections have substantial impacts on animals (Frans and Liu 2024). Many animal species migrate over great distances, thus connecting ecosystems that are otherwise distinct (→ Ecological Connectivity) (Fudickar et al. 2021). Systemic global environmental change is generated by actions that happen locally but change global systems that in turn have effects everywhere, thus linking local actions in one place to changes in other places (→ Global Environmental Change). For example, climate change produces heat waves, drought, increased storm severity, and other weather changes that can stress populations of humans and other animals in places far distant from where the greenhouse gas emissions that drive climate change occurred (→ Climate Change). Given telecoupling, efforts to move toward sustainability must take account of not only local conditions and dynamics but also influences that may be remote from the coupled human and natural system being analyzed (→ Coupled Human and Natural Systems).

Trade is a major force impacting animals across the globe (→ Wildlife Trade). The animal species involved are diverse, from crickets to lions, and they are used as food, in traditional medicine, and as entertainment (→ Insects as Food). The consequences for animal welfare and for the protection of threatened and endangered species are very substantial (→ Welfare) (→ Endangered Species Act) (Lambert et al. 2025). Even when animals are not traded directly, global markets for commodities can have substantial impacts in places far from where the products are consumed. For example, demand for soybeans in Asia leads to intensive farming in the Amazon and in North America (→ Feed Production) that in turn impacts ecosystems and habitats for many animal species (da Silva et al. 2023).

DOI: 10.4324/9781003528142-20

References

da Silva, Ramon Felipe Bicudo, Emilio F. Moran, James D.A. Millington, Andrés Viña, and Jianguo Liu. 2023. "Complex Relationships between Soybean Trade Destination and Tropical Deforestation." *Scientific Reports* 13 (1): 11254. https://doi.org/10.1038/s41598-023-38405-1.

Frans, Veronica F., and Jianguo Liu. 2024. "Gaps and Opportunities in Modelling Human Influence on Species Distributions in the Anthropocene." *Nature Ecology & Evolution* 8 (7): 1365–1377. https://doi.org/10.1038/s41559-024-02435-3.

Fudickar, Adam M., Alex E. Jahn, and Ellen D. Ketterson. 2021. "Animal Migration: An Overview of One of Nature's Great Spectacles." *Annual Review of Ecology, Evolution, and Systematics* 52 (1): 479–497. https://doi.org/10.1146/annurev-ecolsys-012021-031035.

Lambert, Helen, Angie Elwin, Délagnon Assou, et al. 2025. "Chains of Commerce: A Comprehensive Review of Animal Welfare Impacts in the International Wildlife Trade." *Animals* 15 (7): 971. https://doi.org/10.3390/ani15070971.

Liu, Jianguo, Vanessa Hull, Mateus Batistella, et al. 2013. "Framing Sustainability in a Telecoupled World." *Ecology and Society* 18 (2): 26.

Trophic Cascades

Trophic cascades are the indirect species interactions that originate with predators and spread downward through food webs. Since trophic cascades begin with higher trophic level species and proceed through top-down impacts on other species in the food web, they usually can be seen only when the controlling predator is removed from or increases in population in an ecosystem (Ripple et al. 2014). Food-web control through bottom-up forcing is not considered a trophic cascade. The cascade is driven by the effects of predators killing and eating (trophic) other species "otherwise predators would starve and prey would incur no risk of predation" (Ripple et al. 2014, 847). While top-down effects can change the abundance of plants that in turn trigger a change in the food chain, those bottom-up influences are "knock-on effects" of trophic cascades.

The classic example of a trophic cascade is the study by Estes and Palmisano (1974) of the influence of sea otters on marine communities. When sea otters are present, they eat sea urchins who feed on kelp, allowing the kelp to flourish (→ Keystone Species). When otters are absent, the urchin population becomes too large, and their consumption of kelp devastates entire kelp beds (a trophic cascade) that feed other species (a knock-on effect) including bald eagles and gulls and influence the growth rates of mussels and barnacles (→ Animal Forests).

Another well-known example of a trophic cascade is the reintroduction of wolves in Yellowstone National Park in 1995 that led to a cascade of indirect effects on both animals and plants in the Yellowstone ecosystem (→ Keystone Species). Wolves prey on elk, and elk browse on grasses, bushes, saplings, and small trees. In response to the presence of wolves, elk herds must keep moving and cannot remain in place for long (→ Ecology of Fear), so the vegetation is not heavily grazed and has time to recover between migrations. There is also a trophic cascade involving wolves, bears, ungulates, and plants in Yellowstone. The presence of wolves who prey on ungulates reduces the competition between elk and grizzly bears for berry-producing shrubs, providing more available fruit to threatened grizzly bears (Ripple et al. 2014). The top-down influence of wolves to elk to berry-producing shrubs is a trophic cascade, and the consumption of berries by grizzly bears is a knock-on effect (Ripple et al. 2014). Wolves also keep coyote populations in check (→ Mesopredator Release). With the reintroduction of wolves to Yellowstone, coyote populations have been substantially reduced, and their prey of small rodents have rebounded providing food for other carnivores including eagles and hawks.

The loss of large apex predators has had a pervasive influence on the natural world. Their disappearance in marine, terrestrial, and freshwater ecosystems has had cascading effects on processes as diverse as disease, wildfire, carbon sequestration, invasive species, and biogeochemical cycles (Estes et al. 2011). There is an "urgent need for interdisciplinary research to forecast the effects of trophic downgrading on process, function, and resilience in global ecosystems" (Estes et al. 2011, 301).

References

Estes, James A., John Terborgh, Justin S. Brashares, et al. 2011. "Trophic Downgrading of Planet Earth." *Science* 333 (6040): 301–306.

Estes, James A. and John F. Palmisano. 1974. "Sea Otters: Their Role in Structuring Nearshore Communities." *Science* 185: 1058–1060.

Ripple, William J., Robert L. Beschta, Jennifer K. Fortin, and Charles T. Robbins. 2014. "Trophic Cascades from Wolves to Grizzly Bears in Yellowstone." *Journal of Animal Ecology* 83: 223–233.

U

Utilitarianism

Utilitarian ethical theory holds that an action should be judged based on the effect it has on the well-being of sentient creatures (→ Sentience) (Sebo 2023; Frey 2011). The founders of the theory, Jeremy Bentham and John Stuart Mill, argued that since animals are capable of suffering, their well-being must be given ethical consideration (Garner 2018). As Jeremy Bentham put it in 1789, "The question is not, Can they reason? nor, Can they talk? but, Can they suffer?" (Bentham 1789, 143). Peter Singer's *Animal Liberation* facilitated a new wave of discussion on utilitarian analyses of human treatment of other animals (Singer 1975).

The utilitarian approach is usually seen as implying that species may differ in the degree to which they can experience pleasure and pain, focusing on the aggregate well-being of animals rather than on, for example, assigning rights to individual animals or focusing on ethical practices rather than on the outcome of those practices (→ Rights) (→ Virtue Ethics). Recent discussions are extending the ideas of utilitarian animal ethics. Some have suggested that utilitarian arguments might be applied to artificial intelligence (AI) if it becomes sentient and that ethical systems developed for AI should consider their well-being (→ Artificial Intelligence). While utilitarianism seems similar to capabilities ethics, Nussbaum emphasizes that utilitarian analysis focuses on suffering while the capabilities approach emphasizes flourishing and thus is broader in its consideration of what is ethical (→ Capabilities Ethics) (Nussbaum 2022).

Utilitarianism underpins formal economic analysis with the assumption that individuals seek to maximize their individual well-being, usually called utility. This economic logic is the basis of cost-benefit analysis, one of the fundamental tools of economic policy analysis. Most cost-benefit analyses are anthropocentric (→ Anthropocentrism) but recent efforts incorporate consideration of animal well-being (Budolfson et al. 2024). This is part of a general movement to make utilitarian economic analyses better suited to considerations of sustainability (Ekardt 2022; Organization for Economic Cooperation and Development 2018).

DOI: 10.4324/9781003528142-21

References

Bentham, Jeremy. 1789/1907. *An Introduction to the Principles of Morals and Legislation*. Clarendon Press. https://oll.libertyfund.org/titles/bentham-an-introduction-to-the-principles-of-morals-and-legislation.

Budolfson, Mark, Romain Espinosa, Bob Fischer, and Nicolas Treich. 2024. "Monetizing Animal Welfare Impacts for Benefit – Cost Analysis." *Journal of Benefit-Cost Analysis* 15 (S1): 206–223. https://doi.org/10.1017/bca.2024.19.

Ekardt, Felix. 2022. *Economic Evaluation, Cost-Benefit Analysis, Economic Ethics: A Review with Regard to Climate Change – Figures in the Sustainability Discourse*. Springer International Publishing.

Frey, R.G. 2011. "Utilitarianism and Animals." In *The Oxford Handbook of Animal Ethics*, edited by Tom L. Beauchamp and R.G. Frey, 172–197. Oxford University Press.

Garner, Robert. 2018. "The Contemporary Debate in Animal Ethics." In *Animals: A History*, edited by Peter Adamson and G. Fay Edwards, 319–344. Oxford Philosophical Concepts. Oxford Academic. https://doi.org/10.1093/oso/9780199375967.003.0019.

Nussbaum, Martha C. 2022. *Justice for Animals: Our Collective Responsibility*. Simon & Schuster.

Organization for Economic Cooperation and Development. 2018. *Cost-Benefit Analysis and the Environment: Further Developments and Policy Use*. OECD Publishing. https://doi.org/10.1787/9789264085169-en.

Sebo, Jeff. 2023. "Utilitarianism and Nonhuman Animals." In *An Introduction to Utilitarianism*, edited by R.Y. Chappell, D. Meissner, and W. MacAskill. Hackett Publishing. https://utilitarianism.net/guest-essays/utilitarianism-and-nonhuman-animals/.

Singer, Peter. 1975. *Animal Liberation*. Avon.

V

Values

The term values has several related meanings that are central to thinking about animals and sustainability (Dietz 2023; Kalof and Satterfield 2005; Dietz et al. 2005). In everyday language, values refer simply to things that someone considers important. The definition of values in social psychology makes this idea more precise, defining values as "the principles or moral standards of a person or social groups, the generally accepted or personally held judgment of what is valuable and important in life" (Dietz et al. 2005, 339). Most research identifies six values that are important to decisions about sustainability, the environment, and other species: self-interest (→ Anthropocentrism), altruism toward other humans, altruism toward non-human species, and the biosphere (→ Altruism), openness to change, traditionalism, and hedonism. Some research also suggests that concern for non-human animals is a distinct value (Dietz et al. 2017; Schmid et al. 2025). These values, especially altruism toward the biosphere, are found to be important influences on sustainability decisions in many studies (Steg 2023). When the theory was first proposed, feminist theory was invoked to suggest that women and minorities might be higher in altruism than dominant groups in society, and that the values of white men in US society may be "atypical" (Stern et al. 1993). Subsequent studies have often found substantial value differences across gender, race, and class lines (Bouman et al. 2021; Kalof et al. 2002).

In economics and policy analysis, the term value refers to the importance an individual or society assigns to a good or service or state of the world. Economic analysis uses prices as indicators of these values, based on an argument derived from utilitarianism and economic theory that prices reflect the expression of preferences by individuals (→ Utilitarianism). Consideration of sustainability and the well-being of humans and other animals enter into this type of value to the extent that such concerns are part of an individual's preferences. In a sense, the social psychological approach to values provides a theory of why people have the preferences they have. Using such economic values for decision making usually gives those with the most wealth the most

DOI: 10.4324/9781003528142-22

influence and thus contrasts with ideas of ecodemocracy (→ Ecodemocracy) (Dietz and Whitley 2018).

In philosophy, the term values is used in a general sense to indicate what importance is, or should be, assigned to various considerations in making decisions. One of the key journals in environmental philosophy is called "Environmental Values" an indication of the broad meaning of the term. The social psychological and economic approaches are sometimes invoked in philosophical discussion of values, but in philosophy the term values covers a variety of ethical arguments that are not easily mapped onto these definitions of values. In general, the term values is central to thinking about human interactions with other species and sustainability, but care must be taken when the term is used to be clear about what is meant in a particular context.

References

Bouman, Thijs, Ellen van der Werff, Goda Perlaviciute, and Linda Steg. 2021. "Environmental Values and Identities at the Personal and Group Level." *Current Opinion in Behavioral Sciences* 42: 47–53.

Dietz, Thomas. 2023. *Decisions for Sustainability: Facts and Values*. Cambridge University Press.

Dietz, Thomas, Amy Fitzgerald, and Rachel Shwom. 2005. "Environmental Values." *Annual Review of Environment and Resources* 30: 335–372.

Dietz, Thomas, and Cameron T. Whitley. 2018. "Inequality, Decisions, and Altruism." *Sociology of Development* 4 (3): 282–303.

Dietz, Thomas, Summer Allen, and Aaron McCright. 2017. "Integrating Concern for Animals into Personal Values." *Anthrozoös* 30 (1): 109–122.

Kalof, Linda, and Terre Satterfield, eds. 2005. *The Earthscan Reader in Environmental Values*. Earthscan.

Kalof, Linda, Thomas Dietz, Gregory A. Guagnano, and Paul C. Stern. 2002. "Race, Gender and Environmentalism: The Atypical Values and Beliefs of White Men." *Race, Gender & Class* 9 (2): 1–19.

Schmid, Désirée F., Tobias Brosch, and Nadja Contzen. 2025. "Towards a Refined Value Typology for Environmental Psychology." *Journal of Environmental Psychology* 106: 102740. https://doi.org/10.1016/j.jenvp.2025.102740.

Steg, Linda. 2023. "Psychology of Climate Change." *Annual Review of Psychology* 74 (1): 391–421. https://doi.org/10.1146/annurev-psych-032720-042905.

Stern, Paul C., Thomas Dietz, and Linda Kalof. 1993. "Value Orientations, Gender and Environmental Concern." *Environment and Behavior* 25 (3): 322–348.

Veganism

Veganism can be understood as a way of living, a moral framework, and a movement, although people engage with it to varying degrees and for different reasons. Broadly, vegans do not consume non-human animal flesh or animal products. Motivations to be vegan tend to fall into three overlapping categories: environmental, dietary, and ethical.

Environmental vegans avoid consuming non-human animal flesh and products to address land use issues, limit biodiversity loss, and reduce greenhouse-gas emissions (→ Biodiversity) (→ Climate Change). Given that industrial animal agriculture generates roughly 16.5 percent of global greenhouse-gas emissions (→ Livestock Production) (→ Feed Production) (Twine 2021), veganism is increasingly framed as a sustainability strategy. Shifting toward plant-based eating can substantially reduce ecological impacts. For instance, in a global analysis of 38,000 farms, Poore and Nemecek (2018) suggest that vegan diets could lower individual food-related emissions by up to 73 percent and land use by 76 percent.

Dietary vegans abstain from eating non-human animal flesh, dairy, and eggs because of real or perceived health impacts. Positive outcomes of vegan diets include drops in cardiovascular disease, reductions in some cancers, and lower rates of obesity. However, nutritional deficiencies in vitamin B12 may enhance the prevalence of anemia, neurological issues, and bone health peculiarities (Clemente-Suárez et al. 2025).

Ethical vegans extend their commitments beyond diet to encompass a broader refusal to consume or purchase goods derived from practices they understand as exploiting non-human animals (Alvaro 2017). This typically includes avoiding products such as leather, wool, silk, and fur, as well as cosmetics and personal care products tested on animals (→ Animal Research). It may also involve abstaining from supporting entertainment industries that use animals or, for some individuals, questioning the ethics of keeping companion animals.

Veganism as a social movement not only challenges the practice of eating animals (→ Carnism), but is also interested in the ethics of consumption, the structure of global food systems, environmental justice, and the sustainability of life on the planet (→ Environmental Justice). Veganism sits at the intersection of ethics, ecology, and human-animal relations with some utilitarian philosophers arguing that vegetarian or vegan diets are morally preferable because the significant suffering imposed on sentient animals in contemporary agriculture far exceeds the comparatively minor pleasures humans derive from consuming animal products (→ Utilitarianism) (→ Vegetarianism) (→ Sentience) (Singer, 1973). While early vegan advocacy was rooted in animal rights and philosophical commitments to minimizing harm (→ Advocacy) (→ Rights) (Singer 1973; Regan and Singer 1989), contemporary vegan activism has broadened to address intersections of sustainability and to address climate change, biodiversity loss, public health issues, and environmental justice (Wright 2021; Cherry 2006). However, vegan activism is not without internal debates with tensions rising over inclusivity, intersectionality, and commercialization (→ Intersectional Environmentalism) (Twine 2017).

While individuals adopt vegan practices for environmental, dietary, or moral reasons, the movement increasingly focuses on sustainability and justice. Rather than viewing non-human animals as resources, veganism challenges systems that normalize animal exploitation and environmental degradation. Scholars and activists argue that shifting toward plant-based food systems supports healthier ecosystems, reduces pollution, and aligns with principles of compassion and interdependence (Twine 2010; Wright 2021). Although debates remain around access, equity, and cultural diversity, veganism encourages more sustainable and equitable relationships among humans, animals, and the planet.

References

Alvaro, Carlo. 2017. "Ethical Veganism, Virtue, and Greatness of the Soul." *Journal of Agricultural and Environmental Ethics* 30 (6):765–781.

Cherry, Elizabeth. 2006. "Veganism as a Cultural Movement: A Relational Approach." *Social Movement Studies* 5 (2): 155–170.

Clemente-Suárez, Vicente Javier, Laura Redondo-Flórez, et al. 2025. "Impact of Vegan and Vegetarian Diets on Neurological Health: A Critical Review." *Nutrients* 17 (5): 884.

Poore, Joseph, and Thomas Nemecek. 2018. "Reducing Food's Environmental Impacts through Producers and Consumers." *Science* 360 (6392): 987–992.

Regan, Tom, and Peter Singer. 1989. *Animal Rights and Human Obligations*. Prentice Hall Upper Saddle River.

Singer, Peter. 1973. *Animal Rights*. Springer.

Twine, Richard. 2010. *Animals as Biotechnology: Ethics, Sustainability and Critical Animal Studies*. Routledge.

Twine, Richard. 2017. "A Practice Theory Framework for Understanding Vegan Transition." *Animal Studies Journal* 6 (2): 192–224.

Twine, Richard. 2021. "Emissions from Animal Agriculture — 16.5% Is the New Minimum Figure." *Sustainability* 13 (11): 6276.

Wright, Laura. 2021. *The Routledge Handbook of Vegan Studies*. Routledge.

Vegetarianism

In a broad sense, vegetarianism is the practice of abstaining from eating meat such as beef, pork, chicken, or fish, while also still consuming animal products like dairy, eggs, and honey. There are several different types of vegetarians, such as lacto-ovo or traditional vegetarians (no meat, but eat dairy and eggs), lacto-vegetarians (no meat, eat dairy, but not eggs), ovo-vegetarians (eat eggs, but not dairy), pescatarians, also known as pesco-vegetarians (eat dairy, eggs, and fish, but not other types of meat), and flexitarians (eat dairy and eggs, mostly eat plant-based foods, but includes some meats).

Ovo-vegetarians may choose to exclude dairy products for ethical reasons, particularly because contemporary dairy production often involves the routine separation of cows and their calves shortly after birth and the culling of male calves who cannot be used in profitable milk production (Placzek et al. 2021). These practices can conflict with personal or cultural values regarding animal welfare (→ Welfare). Conversely, lacto-vegetarians may abstain from eating eggs for reasons ranging from ethical concerns about the culling of male chicks in hatcheries to a simple aversion to the taste and texture of eggs.

Pescatarians or pesco-vegetarians maintain a primary plant-based diet that also includes consumption of seafood, but red meat and poultry are excluded. Research suggests that most people who ascribe to a pescatarian diet do so because of health and not because of concern for the environment or non-human animals (Rosenfeld and Tomiyama 2021b). Flexitarians eat mostly plant-based foods but may occasionally eat meat. For many, this is easier to follow and easier to maintain long-term. These individuals may be motivated by personal health or environmental impacts of meat eating but need a less restrictive diet due to cultural norms around eating, because they still enjoy the taste of meat, or because of accessibility and convenience.

Vegetarian practices are embedded in broader social systems, shaped by identity, moral meaning, and cultural norms about food and consumption. Just like veganism, motivations to be vegetarian range from ethical concerns for animal welfare to religious and cultural traditions, health-related goals, and values (→ Veganism) (→ Welfare) (→ Religion) (→ Values). Specifically, individuals holding values more aligned with individuality and independence (Nezlek 2025) and less aligned with tradition (Dietz et al. 1995) are more likely to be vegetarian. Compared to women, men consume more meat and are less open to being vegetarian (Rosenfeld and Tomiyama 2021a). In addition, greater conformity to traditional gender roles, male/masculine and female/feminine, predicts more consumption of beef and chicken and lower openness to being vegetarian (Rosenfeld and Tomiyama 2021a).

Vegetarianism encompasses a range of dietary practices that reduce meat consumption. Because vegetarian diets rely more heavily on plant-based foods, they are generally associated with lower environmental impacts compared to carnivore diets that require more land, water, and energy inputs to produce animals for consumption (→ Carnism) (→ Ecological Footprint). However, vegetarianism is not automatically synonymous with sustainability. Industrial dairy and egg production remain resource-intensive and environmentally burdensome, contributing to pollution while also generating significant harm for both human and non-human animals (→ Livestock Production) (→ Pollution).

References

Dietz, Thomas, Ann Stirling Frisch, Linda Kalof, Paul C. Stern, and Gregory A. Guagnano. 1995. "Values and Vegetarianism: An Exploratory Analysis." *Rural Sociology* 60 (3): 533–542.

Nezlek, John B. 2025. "Rethinking Vegetarianism: Differences between Vegetarians and Non-Vegetarians in the Endorsement of Basic Human Values." *PLoS One* 20 (5): e0323202.

Placzek, Matthias, Inken B. Christoph-Schulz, and Kerstin Barth. 2021. "Public Attitude towards Cow-Calf Separation and Other Common Practices of Calf Rearing in Dairy Farming — A Review." *Organic Agriculture* 11: 41–50.

Rosenfeld, Daniel L., and A. Janet Tomiyama. 2021a. "Gender Differences in Meat Consumption and Openness to Vegetarianism." *Appetite* 166: 105475.

Rosenfeld, Daniel L., and A. Janet Tomiyama. 2021b. "How Proximal Are Pescatarians to Vegetarians? An Investigation of Dietary Identity, Motivation, and Attitudes toward Animals." *Journal of Health Psychology* 26 (5): 713–727.

Virtue Ethics

Virtue ethics argues that ethical choices are those that are based on virtues (Hursthouse and Pettigrove 2023). It has its origins in the work of Aristotle, Confucius, Mencius, and Plato, among others, and thus spans both Western and Eastern philosophy (→ Religion). Generally, a virtue is defined "as a morally good, admirable, or praiseworthy character trait" such as "being benevolent or altruistic, unselfish, fair, responsible, respectful, caring, courageous, honest, just, honorable" (Hursthouse 2011: 125).

Virtue ethics suggest that the human treatment of animals depends not on the characteristics we attribute to animals but in adhering to human practices and virtues that we consider desirable (Abbate 2014; Hursthouse 2011). In contrast, other ethical theories depend on what we assume about animals, such as their ability to feel pain or to have goals and a sense of self (→ Capabilities Ethics) (→ Utilitarianism) (→ Ecofeminism) (→ Sentience). Virtue ethics argues that virtues such as compassion or benevolence should shape our choices. These virtues would help consumers decide whether to eat animal products (→ Vegetarianism) (→ Veganism) (→ Carnism). A virtue-based argument calls for scientists to draw on compassion and humility in making decisions about the use of animals in research (→ Animal Research) (Walker 2025).

Environmental virtue ethics suggest that sustainability is best achieved by humans deploying virtues such as compassion, humility, and care for future generations (→ Empathy) (→ Altruism) (Karagjozi 2025; Dzwonkowska 2024). A logical extension is to expand the scope of these virtues to consideration not just of other humans and future generations, but also to non-human animals. Thus, a virtue ethics approach could be applied to any decision that impinges on non-human animals.

References

Abbate, Cheryl. 2014. "Virtues and Animals: A Minimally Decent Ethic for Practical Living in a Non-Ideal World." *Journal of Agricultural and Environmental Ethics* 27 (6): 909–929. https://doi.org/10.1007/s10806-014-9505-z.

Dzwonkowska, Dominika. 2024. *Virtue Ethics and the Environment*, 1st ed. Routledge. https://doi.org/10.4324/9781003433156.

Hursthouse, Rosalind. 2011. "Virtue Ethics and the Treatment of Animals." In *The Oxford Handbook of Animal Ethics*, edited by Tom L. Beauchamp and R.G. Frey, 119–143. Oxford University Press.

Hursthouse, Rosalind, and Glen Pettigrove. 2023. "Virtue Ethics." In *The Stanford Encyclopedia of Philosophy*, Fall 2023 ed., edited by Edward N. Zalta and Uri Nodelman. Metaphysics Research Lab, Stanford University. https://plato.stanford.edu/archives/fall2023/entries/ethics-virtue/.

Karagjozi, Blerina. 2025. "Virtue, Responsibility, and Sustainability: A Kantian Perspective on Environmental Ethics and Intergenerational Justice." *Academic Journal of Business, Administration, Law and Social Sciences* 11 (1): 98–110. https://doi.org/10.2478/ajbals-2025-0008.

Walker, Rebecca L. 2025. *Of Mice and Primates: Virtue Ethics and Animal Research*, 1st ed. Oxford University Press. https://doi.org/10.1093/9780190089986.001.0001.

W

Welfare

Broadly, welfare refers to the physical, psychological, and social conditions that allow an individual animal (human or non-human) to live a healthy and flourishing life. Animal welfare is a term used to describe the experience of a non-human animal in accessing resources that allow them to express behaviors natural for their species. This may include access to adequate food, water, rest, pain relief, and social interaction (Fraser 2023). Human and non-human animal welfare is a multidimensional framework and cannot be reduced to simple measures such as survival or productivity. It requires the consideration of an animal's species, individuality, subjective experience, and ability to exercise some choice (→ Agency).

Clear examples of the importance of considering non-human animal welfare can be found in livestock production and non-human animal-related research (→ Livestock Production) (→ Animal Research). Studies on livestock production show that production environments designed for efficiency, such as overcrowding barns, speeding up conveyer belts, and the utilization of battery cages, increase stress hormones among non-human animals that can lead to abnormal behaviors and health problems (Broom 2010). In contrast, enriched environments that provide non-human animals with adequate space, stimulation, and social opportunities improve both physiological health and behavior expression. Similar findings emerge in zoos and aquariums, where welfare assessments now examine behavioral diversity, cognitive engagement, and species-specific needs rather than relying only on traditional indicators like disease or mortality (→ Zoos and Aquariums) (→ Cognition) (Whitham and Wielebnowski 2013). Research also highlights how updated welfare indicators can be integrated into wildlife rehabilitation and conservation programs to evaluate long-term well-being (Willette et al. 2023; Artigas and Ruiz 2025). These developments demonstrate how the science around welfare helps institutions and policymakers better understand what animals require to thrive.

Welfare plays an important role in sustainability because long-term ecological and food systems cannot be considered sustainable if they depend on practices that cause widespread suffering or degrade the environment.

DOI: 10.4324/9781003528142-23

However, animal welfare can both complement and conflict with traditional management values and practices (Lynch et al. 2025). Welfare-oriented farming systems often align with environmental goals by reducing overcrowding, promoting healthier animals, and decreasing reliance on antibiotics and intensive resource inputs (Broom 2010). Conservation programs that prioritize welfare, such as habitat protection, species-appropriate enrichment, and humane wildlife management, help ensure that animals are not merely surviving by living in conditions that support ecological stability (→ Conservation Biology). By integrating welfare principles into agriculture, conservation, and policy, societies can work toward sustainability models that value the well-being of both humans and non-human animals.

References

Artigas, Florencia Astorga, and Cristian Ugaz Ruiz. 2025. "Suitability Assessment Protocol for the Release of Raptors for Rehabilitation: Based on the Delphi Method of Animal Focused Welfare Indicators." *Open Veterinary Journal* 15 (10): 5078.

Broom, Donald M. 2010. "Animal Welfare: An Aspect of Care, Sustainability, and Food Quality Required by the Public." *Journal of Veterinary Medical Education* 37 (1): 83–88.

Fraser, David. 2023. *Understanding Animal Welfare: The Science in Its Cultural Context*. Wiley.

Lynch, Kate E., Benjamin L. Allen, Oded Berger-Tal, et al. 2025. "Explicit Value Trade-Offs in Conservation: Integrating Animal Welfare." *Trends in Ecology & Evolution* 40 (6): 593–600.

Whitham, Jessica C., and Nadja Wielebnowski. 2013. "New Directions for Zoo Animal Welfare Science." *Applied Animal Behaviour Science* 147 (3–4): 247–260.

Willette, Michelle, Nicki Rosenhagen, Gail Buhl, Charles Innis, and Jeff Boehm. 2023. "Interrupted Lives: Welfare Considerations in Wildlife Rehabilitation." *Animals* 13 (11): 1836.

Wilderness Animals

Wilderness (or truly wild) animals live in remote habitats and try to avoid humans (Donaldson and Kymlicka 2011). But even in remote habitats, human activities have significantly impacted their lives through hunting, capture, management schemes, and the destruction or colonization of their territories and resources (Donaldson and Kymlicka 2011).

Biopolitics, structures that manage and control animal populations, and activities like overhunting, trophy hunting, and wasteful practices such as poaching and unregulated harvesting have caused the extinction of many large animal species (→ Biopolitics) (→ Wildlife Trade) (→ Habitat Loss) (→ Defaunation) (→ Empty Forest) (Kalof and Whitley 2021). By the close of the last Ice Age, a combination of hunting and environmental shifts led to the disappearance of giant wombats, large ground sloths, mammoths, mastodons, cave bears, rhinoceros in Europe, horses in the Americas, pygmy elephants and hippopotamuses on Mediterranean islands, and the marsupial lion in Australia (→ Extinction) (Clutton-Brock 1999; Kalof 2007). However, the scale of ancient hunting is far exceeded by modern human impacts on ecosystems. Today's consumption patterns not only directly affect animals but also alter ecosystems and sometimes even the entire biosphere (→ Keystone Species) (Brashares and Gaynor 2017, 136). The demand for wild meat and the cultural significance attached to hunting are intertwined with economic, sociopolitical, and both local and global market factors (Brashares and Gaynor 2017). According to Benítez-López et al. (2017), a meta-analysis revealed that in tropical areas where hunting occurred, mammal populations were reduced by 83 percent and bird populations by 58 percent.

The removal of apex predators due to hunting and habitat loss creates widespread impacts, including trophic cascades that alter animal populations, ecosystems, and even human health (→ Keystone Species) (→ Trophic Cascades) (→ One Health). For instance, fewer lions and leopards in Africa have led to increased baboon populations seeking food near humans, resulting in higher rates of intestinal parasites for both species (→ Zoonosis). Estes et al. (2011) note that many ecological disruptions such as pandemics and ecosystem shifts are caused by losing native top predators or introducing new ones (→ Native Species). In addition, restoring apex predators benefits both ecosystems and nearby human communities. For instance, Prowse et al. (2015) found through computer simulations that dingoes in Australia control native herbivores and provide ecological and financial advantages for cattle rangelands.

Human-caused global environmental changes have influenced wild animal populations by altering migration routes and shifting the timing of seasonal behaviors such as ovulation and earlier reproduction (→ Global Environmental Change) (Kalof and Whitley 2021). Climate change has notable impacts on wildlife (→ Climate Change). According to Palmer (2016), as ocean temperatures rise, species who prefer warm water such as the common dolphin are spreading, while those who inhabit cold water like the white-beaked dolphin are diminishing. Additionally, shrinking Arctic sea ice has led to lower ovulation rates and body condition in female ringed seals decreased survival among Beaufort Sea polar bears, and allowed bowhead whales to migrate into newly ice-free northern regions (Palmer 2016, 135). Rising sea levels are also causing freshwater wetlands to become salty, which has forced species who depend on freshwater, including magpie geese, barramundi, and turtles to abandon places like Kakadu National Park in Australia (Palmer 2016, 135–136).

Human population growth and environmental changes have led to habitat loss for wild animals. Pfeifer et al. (2017) found that 85 percent of forest vertebrate species are affected by fragmentation, with forest edges altering ecological communities (→ Ecological Connectivity) (→ Habitat Loss) (→ Empty Forest). Threatened species such as the Sunda pangolin, Bahia tapaculo, long-billed black cockatoo, and Baird's tapir are negatively impacted by edges, while invasive species such as the green iguana and the common boa constrictor benefit from edges (→ Ecotones) (→ Invasive Animal Species). The study concludes that fewer than half of Earth's forests are unaffected by edge effects due to human activities (Pfeifer et al. 2017).

References

Benítez-López, Ana, Rob Alkemade, Aafke M. Schipper, et al. 2017. "The Impact of Hunting on Tropical Mammal and Bird Populations. *Science* 356 (6334): 180–183.

Brashares, Justin S., and Kaitlyn M. Gaynor. 2017. "Eating Ecosystems: Wildlife Harvest and Depletion Compromise Socioecological Stability." *Science* 356 (6334): 136–137.

Clutton-Brock, Juliet. 1999. *A Natural History of Domesticated Mammals*. Cambridge University Press.

Donaldson, Sue, and Will Kymlicka. 2011. *Zoopolis: A Political Theory of Animal Rights*. Oxford University Press.

Estes, James A., John Terborgh, Justin S. Brashares, et al. 2011. "Trophic Downgrading of Planet Earth." *Science* 333: 301–306.

Kalof, Linda. 2007. *Looking at Animals in Human History*. Reaktion.

Kalof, Linda, and Cameron T. Whitley. 2021. "Animals in Environmental Sociology." In *International Handbook of Environmental Sociology*, edited by Beth Schaefer Caniglia, Andrew Jorgenson, Stephanie A. Malin, Lori Peek, David N. Pellow, and Xiaorui Huang, 289–314. Springer.

Palmer, Clare. 2016. "Climate Change, Ethics, and the Wildness of Wild Animals." In *Animal Ethics in the Age of Humans: Blurring Boundaries in Human-Animal Relationships,* edited by Bernice Bovenkerk and F.W. Jozef Keulartz, 131–150. Springer.

Pfeifer, Marion, Véronique Lefebvre, Carlos A. Peres, et al. 2017. "Creation of Forest Edges Has a Global Impact on Forest Vertebrates." *Nature* 551: 187–191.

Prowse, Thomas A.A., Christopher N. Johnson, Phillip Cassey, Corey J.A. Bradshaw, and Barry W. Brook. 2015. "Ecological and Economic Benefits to Cattle Rangelands of Restoring an Apex Predator." *Journal of Applied Ecology* 52 (2): 455–466.

Wildlife Crime

Wildlife crime is defined as "any breach of national, regional, or international legislation that protect wildlife species" (World Wildlife Federation 2020). Many activities can fall under this definition. Hunting, trapping, and poisoning wildlife out of designated seasons, without required licenses or permits, in refuges and protected areas, with prohibited techniques and equipment, or targeting species, sizes, or age classes that are off-limits are all wildlife crimes. Many jurisdictions have laws restricting holding wildlife in captivity. Laws and regulations on commercial fishing restrict fishing areas and require permits and licenses, set harvest quotas, restrict gear and techniques, and prohibit taking protected species and size classes. In many jurisdictions the destruction of habitat is also prohibited, as is the harassment of marine mammals

(→ Endangered Species Act) (→ Marine Mammal Protection Act). Substantial law enforcement resources at the national and local levels are devoted to policing these crimes (→ Conservation and Green Criminology).

Most international attention is focused on the illegal trade in wildlife (→ Wildlife Trade) that is addressed by the Convention on International Trade in Endangered Species and the corresponding national laws such as the US Eliminate, Neutralize, and Disrupt Wildlife Trafficking Act (16 U.S.C. §§ 7601–7644), Marine Mammal Protection Act, and Endangered Species Act (→ Convention on International Trade in Endangered Species) (→ Marine Mammal Protection Act) (→ Endangered Species Act). Illegal trade involves live animals but also animal products such as horns, ivory, feathers, organs, skins and scales, and meat. These products are sought for a wide variety of purposes, including collectibles, food, and traditional medicine. It is difficult to estimate the magnitude of the trade since data come primarily from seizures of illegal products by law enforcement and thus depend both on what crimes are committed and what law enforcement detects (United Nations Office on Drugs and Crime 2024). A credible estimate puts the global value of the illegal wildlife trade at as much as 35 billion per year (Keskin et al. 2023).

The illegal trade in wildlife has many adverse effects. It is a major stressor in some ecosystems and a serious threat to several endangered species. The trade often moves across international borders and is considered a form of trafficking (→ Telecoupling). It is frequently linked to organized crime and is used to finance arms, drug and human trafficking, and the corruption of government officials. It is also an important source of zoonotic disease (→ Zoonosis) (Aguirre et al. 2021). The trade involves complex wildlife trafficking networks that operate across national boundaries and involve both legal and illegal activities, presenting challenges for intervention (Gore et al. 2023). Extensive efforts are underway to try to curb the illegal wildlife trade, and there is evidence that efforts are having some success for the species who have been given the most attention such as elephants, rhinos, and pangolins (United Nations Office on Drugs and Crime 2024). Successful strategies emphasize not only law enforcement but overall sustainability strategies that can improve environmental quality and human well-being in communities where trafficking and other wildlife crime originate.

References

Aguirre, A. Alonso, Meredith L. Gore, Matt Kammer-Kerwick, et al. 2021. "Opportunities for Transdisciplinary Science to Mitigate Biosecurity Risks from the Intersectionality of Illegal Wildlife Trade with Emerging Zoonotic Pathogens." *Frontiers in Ecology and Evolution* 9 (February): 604929. https://doi.org/10.3389/fevo.2021.604929.

Gore, Meredith L., Emily Griffin, Bistra Dilkina, Aaron Ferber, Stanley E. Griffis, Burcu B. Keskin, and John Macdonald. 2023. "Advancing Interdisciplinary Science for Disrupting Wildlife Trafficking Networks." *Proceedings of the National Academy of Sciences* 120 (10): e2208268120.

Keskin, Burcu B., Emily C. Griffin, Jonathan O. Prell, et al. 2023. "Quantitative Investigation of Wildlife Trafficking Supply Chains: A Review." *Omega* 115 (February): 102780. https://doi.org/10.1016/j.omega.2022.102780.
United Nations Office on Drugs and Crime. 2024. World Wildlife Crime Report 2024: Trafficking in Protected Species. United Nations. https://doi.org/10.18356/9789211064582.
World Wildlife Federation. 2020. "Wildlife Trade & Crime." www.wwf.eu/what_we_do/biodiversity/wildlife_crime/.

Wildlife Tourism

Wildlife tourism is a sustainable nature-based tourist activity undertaken to view and/or encounter wildlife in a range of settings, from captive to semi-captive to the wild, encompassing a variety of interactions from passive observation to feeding, and/or touching the species viewed (Newsome et al. 2005, 18–19). There are two distinctive types of wildlife tourism: tourism based *on* nature or tourism based *in* nature (Burns et al. 2011). Wildlife tourism based *on* nature includes experiences with captive or semi-captive wildlife (zoos, aquariums, wildlife parks, sea pens) that take place in out-of-nature settings with a focus on positive outcomes for humans (→ Anthropocentrism). Wildlife tourism *in* nature includes experiences that take place in natural settings (watching wildlife in national parks, migratory routes, breeding sites, feeding/drinking sites) where humans can develop an appreciation for both wildlife and their natural environment (Burns et al. 2011). There are seven principles of wildlife tourism that embrace an ecocentric ethic (→ Ecocentrism): adopting an intrinsic value for wildlife, creating behavior change through moral obligations and reasoning, utilizing the precautionary principle, recognizing the interconnectedness between humans and nature, locating wildlife in nature, and encouraging self-reflexive management (Burns et al. 2011).

Human-wildlife encounters in the wild are considered either consumptive (feeding, fishing, sport hunting) or non-consumptive (photography, viewing). While some consider hunting a way to preserve biological diversity by helping prevent depletion of other native animal species and providing funding for conservation efforts through license fees, others oppose hunting because it reduces the genetic quality of wildlife populations since hunters kill the largest, strongest, and oldest animals (→ Biodiversity) (→ Native Species) (→ Learning for Survival) (→ Defaunation) (Newsome et al. 2005). An underlying principle of wildlife tourism is that it should foster conservation of species, and so the hunting of wildlife as a tourism activity is considered to be inconsistent, unjustified, and unnecessary (Newsome et al. 2005).

There are numerous direct and indirect effects of tourism on wildlife, including motor vehicle accidents, transfer of diseases, and changes in animal behavior (→ Road Ecology) (→ Zoonosis) (Newsome et al. 2005). In popular tourism areas, some species may abandon the use of critical habitat for nesting or foraging, resulting in increased mortality and reduced health, fecundity, and population levels. Poor management of human waste, anthropogenic food availability, and feeding animals can lead to an imbalance in the park ecosystems. Increased contact with humans can reduce wildlife fear of humans, causing problems such as animal foraging of food storage containers, destruction of crops, and human-animal conflict (→ Ecology of Fear) (→ Conflict with Terrestrial Animals). For some species the interaction with humans is stressful, and the Marine Mammal Protection Act regulates such interactions (→ Marine Mammal Protection Act).

Wildlife tourism that operates within an ecocentric ethic contributes to conservation, protection, and sustainability. One example is what Jane Desmond calls "extreme animal tourism" that provides humans with the opportunity to experience wild animals in their natural habitat and become immersed in the animals' world. Wildlife tourism in the Galapagos Islands, for example, is an experience of extreme animal tourism where wildlife is abundant, the human presence is limited, and visitor behavior is closely monitored and mediated by local Ecuadorians since it is impossible to legally visit without a licensed guide (Desmond 2017).

In some regions, wildlife tourism represents over 30 percent of all international travel, growing at rates that outpace other forms of tourism and underscoring the importance of the impacts and ethics of the use of animals (Fennell 2024). Environmental groups have joined the wildlife tourism industry to finance the development of conservation efforts such as using tourist fees to support conservation in wildlife areas and national parks (Newsome et al. 2005). Finally, in an exploration of how wildlife tourism might look in the future, Bertella (2021) argues that with the global loss of forests and increase in urbanization, tourists might only encounter wild animals in captive settings. This would change what qualifies as wild, with cows eating grass in a farm setting considered as extraordinary, exotic, and fascinating as elephants grazing on a savannah.

References

Bertella, Giovanna. 2021. "Introduction: Welcome to the Futures of Wildlife Tourism." In *Wildlife Tourism Futures: Encounters with Wild, Captive and Artificial Animals*, edited by Giovanna Bertella, 1–6. Channel View Publications.

Burns, Georgette Leah, Jim MacBeth, and Susan Moore. 2011. "Should Dingoes Die? Principles for Engaging Ecocentric Ethics in Wildlife Tourism Management." *Journal of Ecotourism* 10 (3): 179–196.

Desmond, Jane. 2017. "Staging Privilege, Proximity, and 'Extreme Animal Tourism'." In *The Oxford Handbook of Animal Studies*, edited by Linda Kalof, 506–524. Oxford University Press.

Fennell, David. 2024. *Tourism and Animal Ethics*, 2nd ed. Routledge.

Newsome, David, Ross K. Dowling, and Susan A. Moore. 2005. *Wildlife Tourism*. Channel View Publications.

Wildlife Trade

A recent report estimates that between 2011 and 2020, 82 million animals who are considered "threatened or endangered" were traded away from their home countries, along with 86 million kilograms of animal products from those species (→ Endangered Species Act) (→ Convention on International Trade in Endangered Species) (CITES Secretariat 2022). However, since a substantial fraction of that trade is conducted illegally, estimates of the magnitude of the problem are uncertain (→ Wildlife Crime) (→ Conservation and Green Criminology). Wildlife trade impacts about 60 percent of all animal species who are considered endangered and threatened. Many of these animals were bred for trade rather than captured from the wild. For example, rhino horns are used in traditional medicine and displayed as a source of prestige, elephants are killed for ivory, pangolins for meat and their scales, parrots and cockatoos are sought as pets, and reptiles are used as pets, meat, and medicine (Donald et al. 2024; Marshall et al. 2025).

The legal and illegal trade in wildlife leads to many sustainability impacts (CITES Secretariat 2022). The massive trade in beaver and other animal pelts had a huge impact on the natural history of North America (Dolin 2010). Trade can lead to overharvesting of targeted species with declines in their population, habitat destruction, and harm to non-targeted animal species (→ Telecoupling). In a study of the impacts of wildlife trade on the decline in populations of mammals, birds, and reptiles, Morton et al. (2021) found that overall species declined in abundance by 62 percent where trade occurs; they conclude that current protective measures fail species even where the harvesting for trade occurs in protected areas. In a review of characteristics known to drive trade in wildlife, Hughes et al. (2023) found that species with larger body sizes, greater abundance, rarity, or with certain valued morphological traits (such as large horns or tusks) are particularly prevalent in trade, driving the risk of extinction for at least 24 percent of terrestrial vertebrates (→ Extinction).

On the other hand, breeding animals for trade can lead to substantial increases in total population of those species, and captive populations might form the basis for restocking wild populations. For example, the population of the American buffalo declined from over 30 million to 325 individuals as a result of hunting and habitat loss, motivated in part by trade but also by a deliberate strategy to expand livestock grazing and to undermine Indigenous people who depended on the buffalo (Duncan and Burns 2023). Now about 30,000 buffalo live in the wild and perhaps 400,000 live in captivity, with many of those in captivity bred for human food (→ Livestock Production) (National Bison Association 2025). Wildlife trade also generates income at least for those involved in the trade, but it is also a cause of zoonotic disease (→ Zoonosis). Evidence regarding the overall impact of the wildlife trade on human well-being, particularly in rural communities, is only beginning to be developed. There is a growing understanding that "(n)early every aspect of wildlife-related commerce and risk mitigation measures has implications for environmental justice ..." (→ Environmental Justice) (Reaser et al. 2025:1).

Wildlife trade has a major impact on many key sustainability issues. Some have proposed improved regulation of the current state of wildlife trade to nurture a more legal, sustainable, safe, humane, and equitable activity. However, others argue that those goals are misleading and unachievable, and "moving towards an end to the commercial trade in wildlife should be our ultimate goal ... a path towards how best to protect wildlife, people, and planet" (→ One Health) (D'Cruze et al. 2021, 1).

References

CITES Secretariat. 2022. *World Wildlife Trade Report 2022*. CITES Secretariat.

D'Cruze, Neil, Jennah Green, Angie Elwin, and Jan Schmidt-Burbach. 2021. "Trading Tactics: Time to Rethink the Global Trade in Wildlife." *Animals* 10 (12): 2456.

Dolin, Eric Jay. 2010. *Fur, Fortune, and Empire: The Epic History of the Fur Trade in America*. W.W. Norton.

Donald, Paul F., Eresha Fernando, Lauren Brown, et al. 2024. "Assessing the Global Prevalence of Wild Birds in Trade." *Conservation Biology* 38 (5): e14350. https://doi.org/10.1111/cobi.14350.

Duncan, Dayton, and Ken Burns. 2023. *Blood Memory: The Tragic Decline and Improbable Resurrection of the American Buffalo*. Alfred A. Knopf.

Hughes, Liam J., Oscar Morton, Brett R. Scheffers, and David P. Edwards. 2023. "The Ecological Drivers and Consequences of Wildlife Trade." *Biological Reviews* 98: 775–791.

Marshall, Benjamin Michael, Aubrey L. Alamshah, Pedro Cardoso, et al. 2025. "The Magnitude of Legal Wildlife Trade and Implications for Species Survival." *Proceedings of the National Academy of Sciences* 122 (2): e2410774121. https://doi.org/10.1073/pnas.2410774121.

Morton, Oscar, Brett R. Scheffers, Torbjørn Haugaasen, and David P. Edwards. 2021. "Impacts of Wildlife Trade on Terrestrial Biodiversity." *Nature Ecology & Evolution* 5: 540–548.

National Bison Association. 2025. "From Plains to Plates: Insights from Bison Data & Stats." Bison by the Numbers, National Bison Association. https://nationalbison.org/bison-by-the-numbers/.

Reaser, Jamie K., Meredith L. Gore, Ines Arroyo-Quiroz, Isla M. Kirkey, Nicholas King, and Gunars Platais. 2025. "Editorial: Advancing the Science of Environmental Justice in the International Wildlife Trade." *Frontiers in Conservation Science* 6 (May): 1613910. https://doi.org/10.3389/fcosc.2025.1613910.

Z

Zoonosis

Zoonosis is defined by the World Health Organization (2020) as any disease or infection that can naturally be transmitted between vertebrate animals and humans, or vice versa, with an emphasis on the two-way transmission of pathogens between humans and animals. Zoonosis also includes diseases that can be transmitted through various routes, such as water, contaminated objects (fomites), or vectors (such as insects). Approximately 61 percent of human pathogens are of zoonotic origin, and more than 70 percent of them originate from wildlife species (→ Wilderness Animals) (Noman et al. 2024). Transmission of zoonotic diseases from animals to humans are facilitated by increasing demand for animal protein; unsustainable agricultural intensification; increased use and exploitation of wildlife; utilization of natural resources accelerated by urbanization, land use change, and extractive industries; travel and transportation; changes in food supply chains; and climate change (→ Carnism) (→ Livestock Production) (→ Feed Production) (→ Wildlife Tourism) (→ Wildlife Trade) (→ Telecoupling) (→ Climate Change) (Randolph et al. 2020).

Past zoonotic global epidemics (pandemics) have caused massive loss of life for humans and other species. For example, zoonotic diseases include the "Black Plague" that killed over a third of Europe's human population, the 1918 flu pandemic that killed over 2 percent of the world's human population, and the COVID-19 pandemic that killed 0.3 percent of the human population (Coulibali et al. 2025).

Biodiversity loss increases the risk of human exposure to zoonosis. When ecological communities lose biodiversity, "the species most likely to disappear are large-bodied species with slower life histories, while smaller-bodied species with fast life histories tend to increase in abundance ... (and) fast-lived species are more likely to transmit zoonotic pathogens" (→ Biodiversity) (Keesing and Ostfeld 2021, 6). Ecological restoration, the removal of invasive species, and the reintroduction of native plants can prevent land use-induced zoonotic diseases (→ Restoration Ecology) (→ Invasive Animal Species) (Reaser et al. 2021).

DOI: 10.4324/9781003528142-24

The transmission of disease and infection from humans to animals (sometimes called reverse zoonosis, zooanthroponoses, or anthroponoses) has recently garnered attention because of its potential devastating effects for non-human species, such as the decline of North American bat populations from white-nose syndrome, a fungus that was likely introduced to caves by humans through contaminated gear or clothing (Ballman et al. 2017). Humans have transmitted the severe acute respiratory syndrome coronavirus 2 (SARS-CoV-2) to a wide variety of mammalian species, including cats, dogs, gorillas, hamsters, lions, mink, tigers, and white-tailed deer. The impact the virus has had on endangered species is thus far unknown because of the cancellation of wildlife research and tourism during the global COVID-19 pandemic (Milich and Morse 2024). Research documents that the 2009 H1N1 influenza pandemic began with human transmission of a closely related virus to pigs (throughout history, human-to-pig transmission of influenza is as common as pig-to-human), and the continued reintroduction of the virus from humans to swine has resulted in increased genetic diversity of swine influenza complicating the annual development of flu vaccines (Milich and Morse 2024).

The transmission of disease between non-human animals is also a pernicious problem, particularly for wildlife. Canine distemper in domestic dogs has contributed to the extinction of the African wild dog in the Serengeti and has caused the reduction of the captive breeding population of wild dogs in Tanzania from 52 to 3 (Sykes et al. 2020). Over the past twenty years, dog-transmitted rabies epidemics have reduced the population of the endangered Ethiopian wolf by 75 percent (Sykes et al. 2020).

Future pandemics are likely to be zoonotic in origin according to the Intergovernmental Science-Policy Platform on Biodiversity and Ecosystem Services (IPBES). Since the connection between different species, including humans, and the environment is the foundation of all life on the planet, "a more holistic world view that moves beyond major drivers of zoonoses to understanding the ecological connectivity in pandemic prevention is a local and global imperative" (Tajudeen et al. 2022, 10). This understanding can be accomplished with a One Health approach to sustainability that addresses the interconnection between humans, animals, and their shared environments to prevent and respond to health threats such as zoonotic diseases (→ One Health).

References

Ballman, Anne E., Miranda R. Torkelson, Elizabeth A. Bohuski, Robin E. Russell, and David S. Blehart. 2017. "Dispersal Hazards of *Pseudogymnoascus destructans* by Bats and Human Activity at Hibernacula in Summer." *Journal of Wildlife Diseases* 53 (4): 725–735.

Coulibali, Estelle, Jules Sadefo Kamdem, and Clovis Foguem. 2025. "A Comparative Analysis of the Socio-Economic and Health Impacts of Three Major Pandemics: The Black Death, Spanish Flu, and Covid-19." *HAL Open Science*. https://hal.science/hal-04977730v1.

Keesing, Felicia, and Richard S. Ostfeld. 2021. "Impacts of Biodiversity and Biodiversity Loss on Zoonotic Diseases." *Proceedings of the National Academy of Sciences* 118 (17): e2023540118. https://doi.org/10.1073/pnas.2023540118.

Milich, Krista, and Stephen S. Morse. 2024. "The Reverse Zoonotic Potential of SARS-CoV-2." *Heliyon* 10: e33040. https://doi.org/10.1016/j.heliyon.2024.e33040.

Noman, Zakaria Al, Shadia Tasnim, Rony Ibne Masud, et al. 2024. "A Systematic Review on Reverse-Zoonosis: Global Impact and Changes in Transmission Patterns." *Journal of Advanced Veterinary and Animal Research* 11 (3): 601–617.

Randolph, Delia Grace, Johannes Refisch, Susan MacMillan, Caradee Yael Wright, et al. 2020. "Preventing the Next Pandemic: Zoonotic Diseases and How to Break the Chain of Transmission." *The United Nations Environment Programme*. Accessed March 19, 2026. https://hdl.handle.net/11288/597494.

Reaser, Jamie K., Arne Witt, Gary M. Tabor, Peter J. Hudson, and Raina K. Plowright. 2021. "Ecological Countermeasures for Preventing Zoonotic Disease Outbreaks: When Ecological Restoration Is a Human Health Imperative." *Restoration Ecology* 29 (4): e13357.

Sykes, Naomi, Piers Beirne, Alexandra Horowitz, Ione Jones, Linda Kalof, et al. 2020. "Humanity's Best Friend: A Dog-Centric Approach to Addressing Global Challenges." *Animals* 10 (502). www.mdpi.com/2076-2615/10/3/502.

Tajudeen, Yusuf Amuda, Habeebullah Jayeola Oladipo, Iyiola Olatunji Oladunjoye, et al. 2022. "Preventing the Next Pandemic through a Planetary Health Approach: A Focus on Key Drivers of Zoonosis." *Challenges* 13 (2): 50. https://doi.org/10.3390/challe13020050.

World Health Organization. 2020. "Zoonoses." www.who.int/news-room/fact-sheets/detail/zoonoses.

Zoos and Aquariums

Zoos and aquariums are institutions that maintain collections of living wild animals for public engagement, scientific research, and conservation (→ Captive Wild Animals) (→ Wildlife Tourism). The origins of zoos and aquariums lie in early menageries and aristocratic collections that prioritized spectacle and status. Over the twentieth century, these institutions became increasingly professionalized through advances in animal husbandry, veterinary care, and environmental education (Minteer et al. 2019). This transition was further established through accreditation bodies such as the Association of Zoos & Aquariums (AZA) and global networks that emphasized conservation and species management. Today, AZA-accredited zoos and aquariums increasingly frame their missions around biodiversity protection and sustainability rather than entertainment (→ Biodiversity) (Patrick et al. 2007; Gusset and Dick 2011; Miranda et al. 2023). This shift reflects both changing public expectations and the growing urgency of biodiversity loss, with many institutions now positioning themselves as active contributors to conservation science and environmental stewardship (→ Sustainability) (Greenwell et al. 2023; Conde et al. 2011). At the same time, scholars continue to note tensions between conservation goals and the ethical challenges associated with captivity (Peng et al. 2025; Braverman 2015, 2012).

In practice, zoos and aquariums engage in a wide range of sustainability-oriented activities that extend beyond their physical sites (Tribe and Booth 2003; Minteer et al. 2019; Miranda et al. 2023). Many accredited institutions participate in coordinated breeding and species survival programs designed to maintain genetic diversity and, in some cases, support reintroduction efforts. These efforts have contributed to the recovery of species such as the California condor and black-footed ferret, illustrating how captive animal conservation can complement habitat-based approaches (Miller et al. 2004; Conde et al. 2011). Similarly, aquariums contribute to marine sustainability through research on coral reef restoration, sustainable fisheries, and marine animal rehabilitation. In addition to biological conservation, these institutions function as sites of public education where visitors encounter environmental narratives, animals, and animal ambassadors, experiences that can shape attitudes and behaviors (→ Animal Ambassadors) (da Silva et al. 2019). Empirical studies suggest that zoo and aquarium visits can enhance conservation knowledge and, in some cases, encourage pro-environmental action (Falk et al. 2007; McNally et al. 2025; Godinez and Fernandez 2019). Increasingly, sustainability is also embedded in institutional operations, including energy-efficient infrastructure, waste reduction initiatives, and sustainable sourcing practices, positioning zoos and aquariums as models for environmentally responsible organizational design (Gusset and Dick 2011; Gill and Warrington 2020).

In the future, zoos and aquariums are likely to play an expanded role in sustainability through research, public engagement and education, and global conservation networks (Miranda et al. 2023). Contemporary scholarship characterizes these institutions as potential "conservation hubs" to connect scientific knowledge production with policy, education, and biodiversity management (Rose et al. 2019). This framing emphasizes not only species preservation but also ecological awareness and ethical responsibility. At the same time, ongoing scrutiny surrounding animal welfare continues to shape practices, prompting innovations in enclosure design, enrichment, and welfare assessment (→ Welfare) (→ Animal-inclusive Design) (Lin et al. 2025; Tallo-Parra et al. 2023; DiVincenti et al. 2023). Zoos and aquariums will likely expand on the use of digital technologies such as AI to promote welfare and engage remote audiences, enhance partnerships with Indigenous and local communities, and increase attention to global issues like climate change, biodiversity loss, and ecosystem restoration (→ Biodiversity) (→ Artificial Intelligence) (→ Indigenous and Decolonial Perspectives) (→ Ecological Restoration) (see e.g., Liptovszky and Polla 2025; Vaccaro and Korach 2025).

As global environmental challenges intensify, zoos and aquariums could occupy a unique role in environmental education to promote conservation and sustainability.

References

Braverman, Irus. 2012. *Zooland: The Institution of Captivity*. Stanford University Press.

Braverman, Irus. 2015. *Wild Life: The Institution of Nature*. Stanford University Press.

Conde, Dalia Amor, Nate Flesness, Fernando Colchero, Owen R. Jones, and Alexander Scheuerlein. 2011. "An Emerging Role of Zoos to Conserve Biodiversity." *Science* 331 (6023): 1390–1391.

da Silva, Rita, Paul Pearce-Kelly, Brian Zimmerman, Meredith Knott, Wendy Foden, and Dalia A. Conde. 2019. "Assessing the Conservation Potential of Fish and Corals in Aquariums Globally." *Journal for Nature Conservation* 48: 1–11.

DiVincenti, Louis, Jr., Allen McDowell, and Elizabeth S. Herrelko. 2023. "Integrating Individual Animal and Population Welfare in Zoos and Aquariums." *Animals* 13 (10): 1577.

Falk, John H., Eric M. Reinhard, Cynthia L. Vernon, Kerry Bronnenkant, Joe E. Heimlich, and Nora L. Deans. 2007. *Why Zoos and Aquariums Matter: Assessing the Impact of a Visit*. Association of Zoos & Aquariums. https://repository.library.noaa.gov/view/noaa/37616.

Gill, Sarena Randall, and Wayne Warrington. 2020. "Sustainability Initiatives in Zoos and Aquariums: Looking in to Reach Out." In *Leisure and Sustainability*, edited by Susan Tirone and Elizabeth Halpenny. Routledge.

Godinez, Andrea M., and Eduardo J. Fernandez. 2019. "What Is the Zoo Experience? How Zoos Impact a Visitor's Behaviors, Perceptions, and Conservation Efforts." *Frontiers in Psychology* 10: 1746.

Greenwell, Phillip J., Lisa M. Riley, Ricardo Lemos de Figueiredo, James E. Brereton, Andrew Mooney, and Paul E. Rose. 2023. "The Societal Value of the Modern Zoo: A Commentary on How Zoos Can Positively Impact on Human Populations Locally and Globally." *Journal of Zoological and Botanical Gardens* 4 (1): 53–69.

Gusset, Markus, and Gerald Dick. 2011. "The Global Reach of Zoos and Aquariums in Visitor Numbers and Conservation Expenditures." *Zoo Biology* 30 (5): 566–569.

Lin, Ga-Yi, Keith Chi Hui Ng, and Eduardo J. Fernandez. 2025. "Animal–Visitor Interactions in Zoos and Aquariums: A Systematic Review." *Animals* 15 (13): 1924.

Liptovszky, Matyas, and Emily Polla. 2025. "Artificial Intelligence's Potential in Zoo Animal Welfare." *Zoo Biology* 44 (5): 403–411.

McNally, Xavier, Thomas L. Webb, Charlotte Smith, Andrew Moss, and Jilly Gibson-Miller. 2025. "A Meta-Analysis of the Effect of Visiting Zoos and Aquariums on Visitors' Conservation Knowledge, Beliefs, and Behavior." *Conservation Biology* 39 (1): e14237.

Miller, Brian, William Conway, Richard P. Reading, et al. 2004. "Evaluating the Conservation Mission of Zoos, Aquariums, Botanical Gardens, and Natural History Museums." *Conservation Biology* 18 (1): 86–93.

Minteer, Ben A., Jane Maienschein, and James P. Collins. 2019. *The Ark and Beyond: The Evolution of Zoo and Aquarium Conservation*. University of Chicago Press.

Miranda, Rafael, Nora Escribano, María Casas, Andrea Pino-del-Carpio, and Ana Villarroya. 2023. "The Role of Zoos and Aquariums in a Changing World." *Annual Review of Animal Biosciences* 11 (1): 287–306.

Patrick, Patricia G., Catherine E. Matthews, David Franklin Ayers, and Sue Dale Tunnicliffe. 2007. "Conservation and Education: Prominent Themes in Zoo Mission Statements." *The Journal of Environmental Education* 38 (3): 53–60.

Peng, S., H.-Mei Yeh, and D.M. Broom. 2025. "Zoos and Other Organisations with Living World Impacts Should Have Honest Policies." *Applied Animal Behaviour Science* 286: 106618.

Rose, Paul E., James E. Brereton, Lewis J. Rowden, Ricardo Lemos de Figueiredo, and Lisa M. Riley. 2019. "What's New from the Zoo? An Analysis of Ten Years of Zoo-Themed Research Output." *Palgrave Communications* 5 (1): 128. https://doi.org/10.1057/s41599-019-0345-3.

Tallo-Parra, Oriol, Marina Salas, and Xavier Manteca. 2023. "Zoo Animal Welfare Assessment: Where Do We Stand?" *Animals* 13 (12): 1966.

Tribe, Andrew, and Rosemary Booth. 2003. "Assessing the Role of Zoos in Wildlife Conservation." *Human Dimensions of Wildlife* 8 (1): 65–74.

Vaccaro, Alexandra, and Jill Korach. 2025. "Expanding the Cultural Reach of Zoos." *Journal of Zoological and Botanical Gardens* 6 (2): 21.

For Product Safety Concerns and Information please contact our EU
representative GPSR@taylorandfrancis.com
Taylor & Francis Verlag GmbH, Kaufingerstraße 24, 80331 München, Germany

www.ingramcontent.com/pod-product-compliance
Ingram Content Group UK Ltd.
Pitfield, Milton Keynes, MK11 3LW, UK
UKHW052218060926
475212UK00006B/181
* 9 7 8 1 0 3 2 8 6 5 6 6 9 *